The Longevity Paradox

ALSO BY STEVEN R. GUNDRY, MD

Dr. Gundry's Diet Evolution

The Plant Paradox

The Plant Paradox Cookbook

The Plant Paradox Quick and Easy

The Longevity Paradox

How to Die Young at a Ripe Old Age

Steven R. Gundry, MD
with Jodi Lipper

This book contains advice and information relating to health care. It should be used to supplement rather than replace the advice of your doctor or another trained health professional. If you know or suspect you have a health problem, it is recommended that you seek your physician's advice before embarking on any medical program or treatment. All efforts have been made to assure the accuracy of the information contained in this book as of the date of publication. This publisher and the author disclaim liability for any medical outcomes that may occur as a result of applying the methods suggested in this book.

HarperCollins books may be purchased for educational, business, or sales promotional use. For information, please e-mail the Special Markets Department at SPsales@harpercollins.com.

FIRST HARPERLUXE EDITION

ISBN: 978-0-06-288817-4

HarperLuxe™ is a trademark of HarperCollins Publishers.

Library of Congress Cataloging-in-Publication Data is available upon request.

19 20 21 22 23 ID/LSC 10 9 8 7 6 5 4 3 2 1

To Edith Morrey, aka "Michelle" in all my books

MAY 5, 1912–APRIL 15, 2018

Thank you for showing me and all readers of this book how to die young at a ripe old age.

You continue to inspire me daily.

Contents

Part III: The Longevity Paradox Program

Introduction: This Is a Test

As I was writing this book, Edith Morrey, whom I identified in all of my previous books as "Michelle," passed quickly and peacefully onward just two weeks shy of her 106th birthday. I first met Edith shortly after I moved my practice from Loma Linda University to Palm Springs, California. When she walked into my examination room, I encountered a thin, tall, erect, beautiful woman with a gorgeous head of hair who was dressed to the nines. To my eyes, she appeared to be about 65 years old. But a quick glance at the chart started my hands shaking. Forget 65 or 75 or even 85. She was in her nineties! The woman standing before me in three-inch heels (I kid you not) appeared shockingly young, yet her chart revealed that she was, in fact, quite old.

In Loma Linda, California—one of the world's best-known "Blue Zones"—I had encountered many healthy centenarians. Nevertheless, I was singularly unprepared to meet Edith. She exemplified a seeming paradox: chronological old age wrapped up in an implausibly youthful, vital physical form.

Edith told me she had attended one of my recent speaking engagements and that I reminded her of someone else she'd heard speak about nutrition more than seventy years ago, when she was just 20 years old. That person was the nutritionist Gayelord Hauser, whose advice she'd been following to the letter ever since. She'd bought and read each of his books, adopted his dietary protocol, and stuck to her guns even when her husbands (she buried two in her time, including one who was a doctor) had told her she was crazy. After a lifetime of following Hauser's advice, she was still fit as a fiddle.

I couldn't believe my good fortune in meeting her. I peppered her with questions, wanting to know more about what exactly she'd learned from Hauser and how she'd maintained her health and vitality for so many years. Though I became her doctor and remained so until the day she passed, I can say for certain that I learned more from her than she learned from me. She showed me that the longevity paradox I'd imagined—the ability to die young at a ripe old age—was indeed a possibility that was available to us all.

As I gained more insight into Edith's (and, by extension, Hauser's) nutritional practices, I took a deep dive into the research of longevity and discovered another paradox that defines aging—that nonhuman ancient genes actually have the power to keep us young. How is this possible? Buckle up—we're about to embark on quite an adventure.

In *The Plant Paradox*, I asked you to hop into an imaginary time machine and take a journey with me back 450 million years, to a time when plants were the only life-forms on Earth. They ruled the land for about 90 million years until insects showed up and started eating them. That was a tough time for the plants, but they were not about to go down without a fight. Plants are amazing organisms that can transform sunlight into matter, a feat we have yet to master. They weren't going to let a little thing like the sudden appearance of millions of tiny predators stop them from growing and reproducing, so they developed complex defense mechanisms to protect themselves. These included chemical compounds that poisoned, paralyzed, or entrapped their predators and others that made them sick or disoriented. In *The Plant Paradox*, I argued that many of the health crises we humans are facing today are a result of unwittingly consuming those plant compounds. (If you haven't read that book, don't worry—it's not a prerequisite to reading and understanding anything and everything that follows here!)

Now I want you to climb into a new time machine with me and go back even further, to a time long before even plants existed: about 3 billion years ago.

We would find ourselves in a vast empty space where the only living things are bacteria and other single cells that can grow and divide without oxygen. In fact, as hard as it is to believe, oxygen is often deadly to these single-celled organisms. As a prelude to something you'll read more about later, these organisms could thrive on what we consider a toxic gas called hydrogen sulfide. But something important is happening in the atmosphere: oxygen levels are rising. These bacteria evolved in an anaerobic (oxygen-free) environment. To them, oxygen is lethal, and the world is suddenly becoming a very dangerous place.

Like all living things, these bacteria, which belong to the class of organisms called prokaryotes, have a biological imperative to survive and pass on their DNA, so they come up with a very clever plan to protect themselves in this hostile new environment. They hop inside other single-celled organisms and make a deal that will dramatically change the course of life on Earth. In exchange for food and a stable, protected home, the bacteria will give their host cell extra energy to fuel its functions and survival. It was that arrangement that resulted in the advanced cells, called eukaryotes, which constitute the cells of algae, fungi, plants, and all animals, including you and me.

Now let's get back into our time machine and zoom forward to the twenty-first century. What if I told you that in the present day, these engulfed bacteria still reside in your cells? As the saying goes, truth is often stranger than fiction. The engulfed bacteria are called mitochondria, and their job is to use the oxygen you breathe and the calories you eat to create energy for all of your cells. But not every type of bacteria made the same deal with those single cells all those billions of years ago. So what happened to the rest of them? As the bacteria in the cells created energy and enabled them to evolve into increasingly more complex creatures, oxygen levels in the atmosphere continued to rise. The remaining bacteria escaped the deadly oxygen by moving into the animals' colons, which resembled the anaerobic environment in which they'd thrived for billions of years.

Would it be too "out there" to suggest that bacteria actually created animals, including humans, so they could avoid oxygen and live safely on Earth? And talk about "out there"; suppose I told you that the bacteria in your gut keep in close contact with their relatives, the mitochondria inside your cells, in order to communicate with them about how things are going on the "other side"? We'll discuss all of this and more in the pages that follow.

Now, you may be asking yourself, what does any of this have to do with longevity? In a word, everything. Because as your bacteria's home, what happens to you depends on what happens to them. It may be hard at first to accept the fact that your fate is in the hands of the trillions of bacteria that live inside, on, and around you. Here's the thing: you're not actually who you think you are. The *real* you—or, more accurately, the *whole* you—includes all those bacteria, and the "you" that you are familiar with is actually only a small part of the whole. In fact, 90 percent of "your" cells are not actually human cells at all. They are the cells of bacteria, viruses, fungi, and worms that live on you and inside you, commonly referred to as the microbiome, or as you astute *Plant Paradox* readers know, your holobiome.

So your longevity is paradoxically tied to the fate of these ancient organisms—the oldest parts of you have the power to help keep you young. It all goes back to the bacteria's need to survive and pass on their DNA. Your body is essentially a condominium for your microbiome or, as I like to call them, your bugs. You are their home. And as you'll soon learn, if you provide a nice, hospitable home for them, they will be exceptionally good tenants. They'll keep the utilities running efficiently, the plumbing in tip-top shape, and even the exterior paint fresh. On the other hand, if you feed them foods they do not thrive on, allow squatters to move in and take over, and let the foundation rot, they'll give up and let the rest of you decay right along with it. Our relationship with our bugs has always been, and continues to be, symbiotic; in other words, their health is dependent on you and vice versa. You take care of them, and they'll take care of you—for the long term.

In addition to being made up of 90 percent foreign cells, we humans are comprised mostly of foreign genes. Believe it or not, 99 percent of all the genes that make up "you" are bacterial, viral, and protozoal genes, not human genes at all. Humans actually have very few genes, and the ones we do have are virtually identical to those in our primate cousins, chimps and gorillas. In fact, the popcorn you may be eating as you read this (that was a test; you wouldn't dare eat corn after reading *The Plant Paradox*, would you?) has 32,000 genes, while you have a measly 20,000.[1] *How can that be?* you say. *Corn has more genes than we do? We're so much more complex than a dumb plant! Okay, fine, maybe corn's got us beat, but surely we have the most genes of any* animal, *right?* Wrong! The water flea, *Daphnia*, has the most genes of any animal, coming in right behind corn at 31,000.[2]

If we humans have so few genes, how did we become so complex? What makes us different from other animals? In a word: *bacteria.* When humans evolved, our bacteria changed, and it was our bacteria, rather than our genes, that made us human.[3] As shocking as it may seem, most of what has happened to us, and what will continue to happen to us in the future, is determined by the state of the bacteria in our gut, mouth, and skin. So let's stop focusing on taking care of the 1 percent and start paying closer attention to the 99 percent of the genes that make up you.

You might be feeling a little bit uncomfortable with the idea that we don't have full control over our own bodies. But in actuality, the opposite is true. When we learn how to be a good host to our microbes, we can gain a lot of control over how well we will age and how long we will live. Your fate does not lie in your genes at all—it lies in your microbiome, and many of your daily decisions about food and personal care products influence how happy or unhappy they will be in their home. Paradoxically—and here's the point to remember going forward—whatever happened to your parents or grandparents, your Ancestry.com or your 23andMe results contribute very, very little to your fate and longevity. Much more of your fate resides with the trillions of organisms living in and on you.

Those microbes have invested a great deal in their condo—they want it to stay in good shape for a long time. Their survival literally depends on you, and yours on them. We know this from experiments with germ-free mice (which are born without contact with the bacteria that would normally populate their gut), which live shorter lives and are more susceptible to disease than mice with normal bacteria, because without communication from those germs, their immune systems never develop properly.[4] Your microbes (or "gut buddies," as I like to call them) are there to help you. You are in the driver's seat of your health and longevity if, and only if, and when, and only when, you give over your fate to them—the other you, or the unseen you within.

In the following pages, I'm going to give you a complete guide to the care and feeding of your gut buddies. In fact, I'm going to give you a veritable Google Maps tour of the entire neighborhood that your holobiome inhabits. The problem is that, like all neighborhoods, yours has its share of bad guys. And if you've been following a standard Western diet and lifestyle, chances are that those bad guys have taken over. They've broken down the all-important gut border wall that separates your other inhabitants from you, put their own needs first, hijacked the supply lines, and deprived your human cells of food and critical information to keep things humming along. And your poor gut buddies, with danger lurking on the streets, have gone into hiding.

But here's the good news: if you starve out the bad guys and throw the good guys a lifeline, the good guys will reemerge, reinforce the border, and revitalize the neighborhood. What's more, those good bacteria will start asking you for more of what they need to succeed.

Your gut bacteria are not only largely in charge of your health and how well you age—they also influence your behavior. With the mapping of the human microbiome in 2017, we've discovered that complex animals such as humans, who have fewer genes than plants and fleas, have uploaded most of their information processing to what I like to think of as our "bacterial cloud," which has vast computing power over our fate and health. Since this genome has so many genes and divides and reproduces incredibly rapidly, your holobiome has immense power to tell "you," your immune system, and the organelles in your cells how things are going in the outside world. Though the genome of a bacterium is only one-tenth of a human cell's genome, National Institutes of Health (NIH) researchers recently showed that the microbiome contributes 8 million unique genes to human bodies. This means there are 360 times as many bacterial genes as human genes within you or me![5] Since bacteria replicate (divide) so quickly and have that much genetic "computing power," our gut buddies are capable of almost instantaneous information processing and communicating, even to the point of influencing our thoughts and actions.

I've been gratified and a little amused through the years when my junk-food addicts or meat-and-potatoes guys come back to see me after a couple of months on my Longevity Paradox program and tell me that they've started to crave salads and other green things. They are shocked by their own behavior, which is now being remotely manipulated by a new set of microbes, their gut buddies, which are sending a loud and clear message to take care of their home. You will see this principle in action throughout this book: give your gut buddies what they want, and they will return the favor.[6] Best of all, the bad bugs that are responsible for your junk-food cravings will leave the building and finally stop torturing you.

Now, perhaps you're thinking, how is it possible that bad bugs have taken over if humans are living longer, healthier lives than ever before? Not so fast. There are a lot of misconceptions about aging that we will soon address, and the first and foremost is that we're doing better than ever in terms of longevity. Yes, the average life expectancy has increased over the last five decades. In 1960, the average life expectancy of American men was 66.4 years; by 2013, it was a full ten years longer.[7] For women, the average ages were 73.1 and 81.1, respectively. But much of this extended life span can be attributed to the fact that we've developed vaccines, antibiotics, and hygiene protocols to beat the main causes of a shortened average life span, namely infectious diseases that disproportionately affected young children. Perhaps we've come to an end in terms of what modern advances can accomplish. Sadly, life expectancy has now declined for the last three years![8] And never forget, people have lived into very old age since recorded time. One of my favorite documented cases is that of Luigi Cornaro, whose treatise *How to Live 100 Years, or Discourses on the Sober Life* chronicled his 102 years of living in the 1400s and 1500s! (And never fear; his "sober" life included a prescription for a daily ration of 500 milliliters—about two-thirds of a bottle—of red wine.)

Today we're starting to see both a decreasing life span and a horribly reduced health span, the length of time people maintain full function. Most people now see their health begin to decline at age 50.[9] Yet we've gotten very good at extending our life span with a host of medical procedures, drugs, and treatments. So we're living longer, but we're not living better. This, dear reader, is another paradox of aging, and it's probably why you picked up this book. This paradox has become so pervasive that many of us assume we're meant to spend the second half of our lives in a state of steady decline. We think it's somehow normal to be on several prescription medications, undergo invasive surgeries, and require joint replacements. In many cases, we even plan ahead for it—moving our bedrooms to the first floor of our homes as a preemptive measure even when we can still climb the stairs, for example, as if there were some sort of deadline on stair climbing! Tell that to the Sardinians, who regularly live to over 100 and climb the hilly pastures until their final days.

As a heart surgeon, I have done my part to extend the lives of tens of thousands of individuals. I'm proud of the fact that I've helped so many people live longer lives, but I quit my job as professor and head of cardiothoracic surgery at Loma Linda University School of Medicine when I learned that much of what I'd been taught about health and longevity—information that many leading doctors still believe is true—was simply wrong.

For the past nineteen years, I've been treating my patients with a combination of nutritional therapy and conventional medicine, and over and over I've seen incredible results. When my patients treat their gut buddies right, they are able to dramatically increase their life spans. As my patients and regular readers know, I've seen dramatic reversals of diseases that many doctors still believe are irreversible, changes that we can track with sophisticated blood work and that my patients can feel and see. Many of these changes are directly linked to alterations we've made to their gut bacteria.[10,11]

Between the results I've observed in my patients, my analysis of an enormous amount of recent research on the gut biome, and my own studies of the world's longest-lived communities, I now know that your gut bacteria to a great extent influence both how long you'll live and how well you'll live. And with the help of my amazing patients, I've put together a program that will drive out the bad guys and make the good guys feel safe and happy in their home so they'll be compelled to completely revitalize their neighborhood both inside and out.

Some elements of the Longevity Paradox program may be familiar, such as eating lots of certain vegetables and getting the right amounts of exercise and sleep, while others, such as tricking your body into thinking it's winter year-round to stimulate your stem cells and spacing out your meals to "wash" your brain at night, are brand-new. These strategies have helped my patients lower their blood pressure and cholesterol markers, significantly reduce symptoms of arthritis and other joint issues, resolve MS, lupus, and other autoimmune conditions, improve heart health, and slow or reverse the progression of cancer and dementia—not to mention lose weight and look decades younger! And they accomplish this without starving, eating twigs, counting calories, or putting in hours at the gym.

It doesn't matter how old you are, how old you feel, or how sick or healthy you may be right now. As on all of the best home-makeover shows, renovations happen quickly when the people in charge have the right materials and are motivated to get the job done. If you follow my plan, within just a few weeks you'll have more gut buddies and far fewer squatters, and you'll start to see and feel a difference in your energy levels, in your lack of symptoms of many of the most common diseases of aging, on your skin, and on the scale.

So let's start transforming your body into the most desirable oceanfront suite on the market for your gut bacteria. They'll be sure to thank you with a long and healthy life.

I

The Aging Myths

Before we get into exactly how to best take care of your gut buddies, let's take a closer look at how they operate in your body and why they're such a critical component of your health and longevity. While we're at it, we'll clear up some confusion, misinformation, and downright falsehoods about how and why we age.

When it comes to your gut bacteria, you have two priorities. First, you need to make the good ones so happy that they'll want to stick around and keep their home lovely and well cared for and make the bad ones so unhappy that they'll flee the premises for good. This will give you the ideal population and diversity of gut buddies that you need for a long life span and health span. Second, it's just as essential to have a strong gut lining, which I and other researchers refer to as the border or mucosal barrier, to keep those gut buddies where they should be (in your intestinal tract) so they can protect you from foreign invaders and avoid being mistaken for invaders themselves. A strong, impermeable barrier is the key to avoiding many of the diseases we associate with "normal" aging.

Let's start with your gut bacteria: what they are, what they do, and why they are such an important piece of the Longevity Paradox.

Chapter 1
Ancient Genes Control Your Fate

I always assumed that I would age much like my dad. As the years passed, he grew overweight and suffered from heart disease and other complications that we often associate with getting older, such as stiff joints, lack of mobility, and decreased muscle mass. And for a while there, I did age almost exactly like him. For many years, I was obese, suffered with migraines as he had every day, and had such bad arthritis that I had to wear braces on my knees to run. Yet I ran thirty miles a week, went to the gym for an hour every day, and ate what I had been taught was a healthy diet. I was doing everything right—or so I thought—and so I figured that my poor health and rapid aging were clearly a case of bad genes. I shared the same DNA as my dad, I reasoned, so I was fated to be fat and sick just like him.

To put it bluntly, I was dead wrong. Thankfully, I figured it out before either of us was actually dead.

Sure, I had certain things in common with my dad, in particular many of the same health issues that he struggled with. But what I've learned since then is that we shared those health problems not because we shared the same DNA, but because we had similar habits and lived in similar environments, both of which had shaped our holobiomes (the trillions of microbes in our bodies, on our skin, and even in a cloud around us) in remarkably similar ways. And it was those holobiomes and *their* genes that were making us age so quickly, not our human genes.[1]

This may seem hard to believe, but recent research suggests that this phenomenon is very real. A statistical analysis published in *Nature* in 2018[2] revealed that our gut microbiome (the part of the holobiome that lives in our guts) is shaped by many factors and that "host genetics" (that's right, you are merely a host for your gut buddies) plays a relatively minor role in determining your health and longevity. Now read that sentence again and really let it soak in. Your human genes have very little to do with your fate. So when a new patient recounts his or her family history now, I know that what I am really hearing is a summary of his or her family's eating habits and lifestyle. In fact, people who are not genetically related but live together in the same household have strikingly similar gut microbiomes.

Even more compelling, for the individuals in the analysis published in *Nature*, the makeup of an individual's gut bacteria was a better predictor of many health outcomes—including blood glucose level and obesity—than genetics. In other words, you have a better chance of sharing the same health conditions as your roommate or your spouse than your biological parents, and that's not because of luck or coincidence. It's because you have similar gut bugs.

Those gut bugs don't just impact a few health outcomes; they directly influence the health and longevity of every part of your body, from your skin to your hormones to your cellular energy levels. And they play a huge role in determining how well and how long you'll live.

A recent study from the China Institute collected and analyzed gut bacteria from more than 1,000 healthy Chinese participants ranging in age from 3 to over 100. They found that a healthy gut is a key indicator of individuals who live past age 100.[3] The study participants who were over 100 years old possessed gut buddies that were strikingly similar to those of individuals who were up to *seven decades* younger. The centenarians had the gut microbes of a 30-year-old!

Meanwhile, a groundbreaking study in 2017[4] was able to pinpoint for the first time the specific types of bacteria (in the Ruminococcaceae, Lachnospiraceae, and Bacteroidaceae families) that are dominant in healthy people between the ages of 105 and 109 but normally decrease in abundance with age. These particular families of bacteria support health during aging, but most of us lose them as we get older. People who live to be 105 years old, however, hold on to specific gut buddies that keep them young.

Still not convinced that your gut buddies play a key role in determining how well and how long you'll live? Consider this: when researchers take feces from obese rats and feed them to skinny rats, presto change-o, the skinny rats become fat. The reverse is also true: eating a skinny rat's poop makes fat rats thin. Need a human example? In the 1930s, psychiatric patients suffering from severe depression were given laxatives to clean out their colons and then given enemas with fecal matter from people who were not depressed. The result: the patients had a marked improvement in mood.

When I was a medical student at the Medical College of Georgia in the 1970s, many patients who had taken broad-spectrum antibiotics—which, at the time, were considered a new class of drugs—developed a severe colon infection called *Clostridium difficile* (also known as *C. diff.*) colitis. We now know it wasn't a coincidence; the antibiotics wiped out the patients' gut bacteria, leaving them more susceptible to the infection. But the truly amazing thing is that many of those patients were cured by fecal enemas made with the poop of healthy medical students, including yours truly. Little did we know then that the gut buddies in our feces restored those patients' health by driving out the overgrown bad guys, *C. diff.*

So having the right gut buddies can make you thinner and happier and even cure you of deadly disease. Would resetting your microbiome to become younger actually make the rest of you younger, as well? It certainly seems so. But how? For the answers, let's take a look at exactly what your gut buddies do.

GUT BUDDIES AT WORK

Unbeknownst to you, the inhabitants of your microbiome are quite busy day and night. They are involved with regulating major aspects of your immune system, your nervous system, and your hormonal (endocrine) system around the clock. But perhaps their most important role is in supporting your digestive system—your gut buddies digest the foods you eat and manufacture and deliver vitamins, minerals, polyphenols, hormones, and proteins to where they are needed in your body.

For years we had no idea how important the microbiome was for digestion, let alone the manufacture of vitamins and hormones. Now we know that if the bacteria in your gut can't process the food you eat, you don't benefit from the nutrition or the information in that food, no matter how good for you it may be. This is true for all animals. Even a termite cannot "eat" wood; bacteria in its tiny gut actually do the work of digesting the wood and converting it into absorbable compounds. Without them, the termite would starve no matter how much of your home it ate. As a 2016 study on the impact of diet on longevity concluded, "Nutrient uptake depends on your microbiome."[5]

In my practice, I have seen many patients who are deficient in several vitamins, minerals, and proteins when they come to me—not because they aren't consuming them but because their microbiome isn't manufacturing and/or allowing absorption of them. When we drive out the bad bugs and motivate these patients' gut buddies to revitalize the neighborhood, lo and behold, the deficiencies disappear. You can think about it this way: *you are not what you eat; you are what your gut buddies digest.* And they can digest only the specific foods they have evolved to recognize and process for you. On the Longevity Paradox program you're going to start eating for your gut buddies instead of for yourself, and they will repay the favor in spades.

To add a note of levity here, one of the greatest long-lived health gurus of the twentieth century, Jack LaLanne, whom I had the pleasure of knowing, used to give this advice on eating: "If it tastes good, spit it out!" What he was actually saying without knowing it at the time was: eat for them, not for you! And cheer up; the food you are going to eat for them will taste great for you, I promise.

But your gut buddies have plenty of other jobs, too. They keep yeast, or candida (which is a normal inhabitant of every gut), in check and fight against the overgrowth of other harmful microorganisms. They also act as bouncers, manning the door to your gut and actually educating your immune system as to which foods and substances are beneficial (or at worst harmless) and should be let in and which ones may harm you and should be barred from entry. This particular job has become increasingly demanding over the years as our diets have become increasingly complex (more on this later).

Your gut buddies also create the precursors to many important hormones and communicate with the rest of the cells in your body about how life is going down in the gut. How exactly do they do this? Among other things, by signaling directly to your mitochondria, of course! Did you think I'd forgotten about those very savvy ancient dealmakers that hopped into our cells for a better place to live and protection and in exchange do the important job of producing all of the energy we need for our cells? Of course not. In fact, I believe they represent a missing link in most discussions about longevity.

THE SISTERHOOD OF BACTERIA

Your gut buddies and your mitochondria have a lot in common besides the fact that they both evolved from ancient bacteria. While your gut buddies digest the food you eat, your mitochondria break down (or "digest") nutrients in order to produce energy. You can think of this as your cellular digestive system. Is it a coincidence that descendants of bacteria control both types of "digestion"? I don't think so, and hopefully by the time you're done reading this book you won't, either.

You inherited both your gut buddies and your mitochondria from your mother. As a refresher, mitochondria are engulfed bacteria that live in all your cells. They contain their own mitochondrial DNA (genes) that actually divide separately from the rest of the cell's DNA contained in the nucleus. (Didn't you just love high school biology?) All mitochondrial DNA is transmitted through the female egg from mother to child. So all mitochondria and their DNA are female. Likewise, your mother initially populated your gut microbiome at birth with her bacteria as you traveled down the birth canal and were exposed to the bacteria in her vagina.

This elegant system continued even further when your mother nursed you for the first time. Amazingly, her milk contained specific sugar molecules called oligosaccharides and galactooligosaccharides that you couldn't digest but were the exact type of food the gut buddies she'd just passed onto you liked best. In other words, Mom was feeding two infants at once: you and your newly introduced (from her) microbiome. She needed to get the two of you off to a great start—a fact that's not only mind blowing but, as you'll soon learn, mind enriching!

Therefore, I like to think of your gut buddies and your mitochondria, which both came from your mother, as sisters. And like all the best sisters, they are constantly talking to each other. (At least that's the case with my own two daughters.) Your gut buddies report to their sister mitochondria about what's happening in their corner of the condo, and the mitochondria respond to those reports by taking one of several actions.

In addition to creating energy for your cells, mitochondria are in charge of cell signaling, cellular differentiation (what kind of cell a given cell should become), cell death, and cell growth. In other words, the mitochondria decide if a cell should grow rapidly, slowly, or not at all. We will see this again later when we talk about cancer cells.

This means that mitochondria play a crucial role in the aging process. In a study conducted at the University of Alabama at Birmingham, when researchers induced a chromosomal mutation in mice that led to mitochondrial dysfunction, the mice developed wrinkled skin and extensive, visible hair loss within a matter of weeks. When the mitochondrial function was restored, the mice's smooth skin and thick fur returned.[6] Moreover, new research suggests that mitochondrial damage—and whether the damage is repaired—is largely dependent on messages the mitochondria receive from their sisters living in the gut. Are you paying attention yet?

You may be wondering how these messages are sent. Well, I like to jokingly refer to them as "text messages," but they are actually hormonal and chemical signals. We used to believe that the nucleus (also known as the command center) of a cell was in charge of all cellular communication, but you can take the nucleus out of a cell, and that cell will do just fine and respond to information without having a control center. This is because communication takes place at the cell membrane or the mitochondrial membrane, not in the nucleus. You can remove the hot spot and still have Wi-Fi. Your gut buddies produce hormonelike substances and fatty acids that enter your blood or lymphatic circulation and then attach to the cell membrane or a mitochondrial membrane of other cells and exchange information.

It's helpful to look at simple creatures such as worms for evidence of this communication system. The roundworm *C. elegans* is a time-tested (excuse the pun) model for longevity, as its primitive gut behaves (in miniature) just like that of a human. A 2017 study[7] looked at the effects of a particular compound called polysaccharide colanic acid (CA), which the worms' gut bacteria produce. When my colleagues conducting the study gave the worms supplements of CA, they lived longer than their counterparts that were not given the compound.[8] How was that possible? The mitochondria in the roundworms' other cells responded to the signal from CA. The extra CA promoted mitochondrial fission, meaning that the existing mitochondria each divided into two separate mitochondria in order to produce extra energy. The study showed that the makeup of the bacterial microbiota could influence aging in the host organism.

It appears that our gut buddies have a remarkably similar process. When our mitochondria receive a message from their sisters in the gut, they, too, respond in ways that increase the number of mitochondria and improve their function. This is a process called hormesis, best summed up by Nietzsche's famous quote "That which doesn't kill me makes me stronger." You'll learn more about activating hormetic signals later. This is precisely why a major component of the Longevity Paradox program involves fostering the kind of bacteria that will send their sisters longevity messages. They will respond by getting stronger and producing more energy while working more efficiently.

Your gut buddies truly influence every aspect of your health and well-being. If they are happy with how you're treating them, they communicate that to their sisters, and in turn you, by making feel-good hormones such as serotonin.[9] They even protect your arteries from harm.[10] If they are hungry or stressed for any reason, they'll send an alert about that, too. These sisters work together pretty seamlessly. Let's hear it for the sisterhood!

However, this can all change if you do something to drive off the good bugs or let too many bad bugs in. Those bad bugs haven't invested in their home the way your gut buddies have, and they have no interest in taking care of you. They're only out for themselves. And they serve their own needs by hijacking the communication system between your gut buddies and their sisters. This makes you crave the foods that *they* need, which are the sugars, fats, junk foods, and fast foods that leave you overweight, inflamed, sick, tired, and more prone to heart attacks, autoimmune conditions, musculoskeletal problems, Alzheimer's disease, and even cancer. Meanwhile, these bad bugs do nothing to protect you from the resulting damage. If anything, they incite more and more of it until they've destroyed their own neighborhood. As in a good Marvel Avengers movie, the bad guys threatening your world must be stopped.

Want to Live Forever? Look to the Naked Mole Rat

Naked mole rats have garnered a lot of attention in the scientific community for their extreme, and even bewildering, longevity. These small, ugly, hairless rodents don't seem to die of old age. They are not immortal, of course, but when they do die, it appears to be random instead of based on age. On the surface, their longevity is perplexing and just downright weird. In fact, there are lots of weird things about naked mole rats. (Even their appearance. Go ahead, google it.) They can live without oxygen for up to eighteen minutes, almost never get cancer, and on average live about ten to fifteen times as long as other rodents their size.

My colleagues in longevity are hot on the trail of the naked mole rats, trying to uncover their secret. How do they seemingly defy the aging process? Many of us now believe the answer lies in what they eat or, more precisely, what they feed their gut buddies. These rodents live in vast subterranean tunnels, where they eat primarily roots and tubers that are extremely hard to digest. Their microbiome aids with that digestion and in doing so produces compounds that massively extend the life span of these rats compared to their short-lived grain-eating cousins.

Among those compounds is hydrogen sulfide.[11] (Remember that those ancient bacteria lived on this instead of oxygen? Well, your mitochondria can do the same!) This may explain why naked mole rats can go eighteen minutes without oxygen: they use hydrogen sulfide to fuel their mitochondria. And where do they get the hydrogen sulfide? From all the tubers, onion- and garlic-like bulbs, roots, and soil fungi that they eat. The fungi are superb sources of compounds that promote longevity. We'll have a full discussion of these guys soon.

Finally, naked mole rats have extremely high levels of hyaluronic acid, thought to make their bodies extremely flexible for maneuvering in their underground tunnels. And where did they get all that hyaluronic acid? Again, from the tubers. In fact, many long-lived people get very high amounts of hyaluronic acid from their sweet potato– and taro root–based diet. (Care to guess what some of the food choices in The *Longevity Paradox* might be?)

The impact of this diet on longevity is reflected in the naked mole rat microbiome. In 2017, a group of my fellow researchers from Italy, Germany, and Ethiopia looked at the fecal matter of naked mole rats and compared their gut bacteria makeup and diversity to that of humans and other rodents. They found that the microbiota of the naked mole rat was approximately as diverse as a human one and significantly more diverse than that of a wild mouse. This explains why naked mole rats greatly outlive other rodents their size: their gut biome more closely reflects those of animals with far longer life spans. But what's more interesting for you and me is the fact that naked mole rats are the only rodents that eat tubers and roots, which are some of the most important sources of food for gut buddies. You can see where this is going, right?

Also of note, the naked mole rats in the study had an abundance of a specific family of bacteria (Mogibacteriaceae) that are also found in humans with extreme longevity (over 105 years old).[12] Read that again! Super old naked mole rats and super old humans have the same species of gut bugs! Coincidence? Not on your life. Want some more "coincidences"? Naked mole rats also have an extremely low metabolic rate, and in times of drought or famine, their metabolic rate drops by an additional 25 percent. This is another key to their longevity that you will learn to duplicate on the Longevity Paradox program.

GUT BUDDY EVOLUTION

But before we stop the bad bugs (and I promise we will), let's take a look at how and why they have taken over so many of our bodies in the first place. This requires another quick trip in the time machine to a much more recent period of history, only about 40 million years ago. It may be hard to believe, but back then we were tree dwellers. We mainly ate tree leaves and other two-leafed plants (known as dicots), along with the fruit of those plants. The other animals in existence were grazers. Grazing animals ate single-leafed plants (known as monocots), such as grasses and their seeds. Our intestines and the microbes within them evolved very differently from those of grazers, and therefore we were each uniquely able to tolerate the plant compounds, including proteins called lectins, that comprised the majority of our diets. In other words, the microbiome of a grass eater evolved to digest the lectins and other substances in monocots, while a tree-leaf eater evolved to digest the lectins and other substances in dicots.

We know that the longer you are exposed to a compound, the more you become tolerant of it. Think of how allergy shots give you a tiny dose of an allergen until eventually you are no longer allergic to that food or substance. In this case, the time frame necessary for us to gain the ability to tolerate certain compounds wasn't weeks or months; it was millennia. Over the course of 40 million years, the microbes that now call your body home and can easily digest two-leafed plants were passed down from generation to generation.[13,14]

Likewise, the predecessors of cows, sheep, antelope, and other grazers have had millions and millions of years to develop and pass on gut buddies that are capable of handling single-leafed plants. By handling, of course, I mean digesting and eliminating them, and if they can't digest them, then at least signaling to their sister mitochondria and the immune system that all is well, the compounds are familiar and don't pose a threat. After all, they've been encountering them for millions of years. But if your gut buddies didn't evolve to grow accustomed to a particular compound, they will see it as a threat and let their sisters know that trouble is brewing. This is where a great deal of unnecessary and avoidable aging begins.

So what are the compounds in single-leafed plants that are so troublesome for human gut buddies? If you've read *The Plant Paradox*, you already know the answer. If not, don't worry—you're going to get up to speed quickly. The compounds that create the most problems for your gut buddies are called lectins. They are a type of "sticky protein" that plants produce as a defense against being eaten. Remember, just like you (and your gut buddies), plants want to live long enough to pass their DNA along to the next generation. So they produce lectins as part of their survival strategy. Before humans existed and insects were the main predators that plants had to look out for, lectins paralyzed the bugs that ate them—a pretty effective defense mechanism.

Now, we're a lot bigger and stronger than a little

bug—and we have our own defense mechanisms, such as mucus, which you'll read more about later. So we often don't experience any immediate problems after eating lectins. But, importantly, our gut buddies do, and the more lectins we eat, the less happy those gut buddies will be. And as you now know, if your gut buddies are unhappy, they will let their home sink into a state of disrepair. And what that means for you is becoming overweight, tired, achy, and sick.

When we look at mice and rats, which are grain eaters, it's easy to see how their microbiomes have evolved to better handle the lectins in grains as a result of having eaten single-leafed plants and their seeds for millions of years. Compared to humans, rodents have hundreds of times more of an enzyme called protease in their gut, which breaks down lectins and other grain proteins. Like rodents, the longer you have been eating a particular plant lectin, the more time your gut buddies have had to develop ways to defuse them.[15]

Our gut buddies and enzymes don't have the same mechanisms as those of grazing animals, but that wasn't a problem until about ten thousand years ago. That's when we began to cultivate grains and other single-leafed plants. These plants are totally different from the foods that your gut buddies evolved the capacity to eat—the ones that make them happy so they will keep you young.

As I said in *The Plant Paradox*, ten thousand years may seem like a long time, but developing an immunological tolerance to a new lectin within that time frame is like speed dating in terms of evolution. And over the last fifty years or so, things have gotten much, much worse for your gut buddies as humans have largely abandoned traditional methods of eating and preparing lectin-rich foods, such as soaking and fermenting, opting instead for quick, cheap options. We've also taken to eating food that has no basis in nature: genetically modified organism (GMO) foods and meat and dairy products from animals that were fed foods (and drugs) that their gut buddies aren't equipped to digest, either! Our diets have changed more rapidly over the last half century than ever before in history. We now eat far more wheat, corn, and other grains, as well as soybeans—often in the form of processed foods—than unprocessed foods such as leafy greens and other vegetables.[16]

During this same time period our food system has been compromised by an onslaught of herbicides, biocides, drugs, fertilizers, and food additives. And chemicals from personal care products, factory-produced furniture, and household cleaners have invaded our homes. Collectively, exposure to these toxins has thrown our holobiome for a loop. A new study out of the University of Colorado at Boulder shows that when people are out and about, they leave plumes of chemicals behind them, not just from their car tailpipes but also from the products they put onto their hair and skin. During rush-hour traffic, emissions of siloxane, a microbiome-destroying ingredient in shampoos, lotions, and deodorants, are found in comparable levels to vehicle exhaust.[17] Just one more reason to dread your daily commute.

Even our healthy vegetables are not being raised with the help of soil bacteria, which have been wiped out by modern farming practices. Soil levels of zinc and magnesium, which both help prevent diabetes and metabolic syndrome, have dropped significantly.[18] There is simply no way for your gut buddies to catch up and adapt to all of these changes so quickly. And that chemical overload, along with drastic changes to our diet, is sending our gut buddies away in droves and making it possible for the bad guys to take over.

GUT BUDDY POISON

But wait—it gets worse. Over the last fifty to sixty years, we've also seen a number of new drugs and medical technology "advances." Many of these discoveries are helping us live longer, but this has come at the expense of our microbiome. And therein lies another paradox.

In the late 1960s and early 1970s, broad-spectrum antibiotics arrived on the scene. These drugs are unique because they are capable of killing multiple strains of bacteria simultaneously, and—don't get me wrong—they have saved countless lives by eliminating bacteria that cause diseases such as pneumonia and septicemia (blood poisoning). However, to your gut buddies, these antibiotics are basically the equivalent of a bomb going off in the middle of the condo. They wipe out everyone without bothering to target the bad guys. As a result, the bad guys are killed off, but so are most of the good guys, throwing off the delicate balance of your inner bacterial population.

I'm grateful that we have antibiotics for instances when we really need them, but they have become so popular that doctors prescribe them when they're not warranted, often even when their best guess is that the patient is suffering from a virus, which an antibiotic won't cure. Before you accept a prescription and head off to the pharmacy, think about this: every course of broad-spectrum antibiotics you take affects your gut biome for up to two years. Some of the gut buddies that get killed may never feel safe enough to return. Lest I be accused of being overly dramatic, here is some evidence of the devastating effects of antibiotics: studies show that every time you take a course of antibiotics, you increase the likelihood of developing Crohn's disease, diabetes, obesity, or asthma later in life.[19]

But even if you've never filled a prescription for antibiotics (which would greatly surprise me), you've still most likely consumed enough of them to kill off large numbers of your gut buddies. How so? Conventionally raised livestock are fed antibiotics in shocking quantities to prevent them from getting sick and to fatten them up for slaughter, and you consume those drugs when you consume meat, milk, and other animal products. Broad-spectrum antibiotics are also given to pigs, chickens, and other animals because they help them grow faster, larger, and fatter. And guess what? They do the same thing to you by killing off the gut buddies that keep you lean and flexible well into your old age. In fact, obesity in humans is marked by a decrease in bacterial diversity in the gut.[20] Too many antibiotics will make you fat and ready for slaughter. Is it any wonder our culture's version of old age doesn't look so inviting?

I hate to tell you this, but you've also likely consumed plenty of "antibiotics" even if you've been a vegan your entire life. That's because glyphosate, the main ingredient in the herbicide Roundup, patented by the agricultural biochemistry giant Monsanto (now owned by Bayer) as an antibiotic, is sprayed on nearly all GMO crops and on many conventionally grown crops as well.

And glyphosate destroys your microbiome and disrupts the gut border just as if you had swallowed a prescription![21] It's found in the meat and milk of animals that are fed grains and beans, as well as in the crop plants you eat and the products made with them that line grocery store shelves. In other words, your vegan bean or pasta dish is most likely loaded with this antibiotic.

In 2015, the cancer agency of the World Health Organization (WHO) declared glyphosate to be a "probable human carcinogen."[22] As a result, the Organic Consumers Association (OCA) and the Feed the World Project (now the Detox Project) teamed up to offer the public the opportunity to have their urine tested for glyphosate. The response was so overwhelming that they had to suspend testing because their lab wasn't big enough! But the results from the first group of more than a hundred individuals who submitted specimens were staggering—93 percent of the urine samples tested positive for glyphosate. Despite what the manufacturer of Roundup may say, glyphosate is clearly in our food supply in large amounts.

In 2018, researchers at Indiana University and the University of California, San Francisco, teamed up to conduct the first-ever study of its kind, testing the urine of seventy-one pregnant women. The results were shockingly similar: they found detectable levels of glyphosate in 93 percent of the urine samples. That is truly alarming considering that glyphosate exposure during pregnancy is associated with shortened gestational length and lifelong health consequences for the child.[23]

When you consume glyphosate, it doesn't just kill your gut buddies, it also disrupts their ability to produce the essential amino acids tryptophan and phenylalanine, which make up serotonin, the "feel good" hormone, and thyroid hormone. Is it any coincidence that a large percentage of our population is taking antidepressants and thyroid medication?

These are just some of the reasons you'll be avoiding conventional animal products on the Longevity Paradox program. The lectin load and the glyphosate load in grains and beans—which are heavily treated with glyphosate—are a double whammy.

In addition, there are now estrogen-like agents in most of our plastics, scented cosmetics, preservatives, and sunscreens. Exposure to these agents is linked to obesity, diabetes, and other metabolic diseases, as well as reproductive problems, women's hormone-sensitive cancers (such as breast and ovarian cancer), thyroid problems, and impaired development of the brain and neuroendocrine systems.[24] Many of these diseases and issues are associated with "normal" aging but are not really normal at all. Remember, your gut buddies produce the precursors to many hormones. This is a primary way they communicate with their sisters in your cells. So compounds that disrupt your hormonal system are hijacking this ancient and essential line of communication.

For proof of the connection between your microbiome and your endocrine system, look no further than antibacterial chemicals such as triclosan that are found in hand sanitizers, soaps, deodorants, toothpaste, and countless other personal care products. They kill off your gut buddies, act like estrogen in the body, and have been shown to stimulate precancerous cells to continue to proliferate.[25] Your mitochondria decide which cells should live and which should die based on hormonal messages from your gut buddies.[26] When estrogen-like substances hijack those messages, cancer cells can grow unchecked.

There is yet another troubling problem with chemicals such as glyphosate in your food and the other chemicals in many personal care products: they reduce your liver's ability to convert vitamin D to its active form so it can do its job of absorbing calcium and promoting healthy bone growth, which, if blocked, sets the stage for osteoporosis as you age. Vitamin D deficiency is now incredibly common. I have observed that around 80 percent of my patients have low levels of vitamin D in their blood and that men with prostate cancer have particularly low vitamin D levels.[27] (You'll read more later about the safe and effective personal care products your gut buddies will love.)

Before we move on, you need to know about one more enemy of your gut buddies, and this one might hurt: sugar. Bad bugs love simple sugar; it's what they live on. Your gut buddies need complex sugar molecules called polysaccharides ("many sugars") to grow and flourish, but the bad bugs thrive on the simple sugar you put in your mouth every day. This is one of the main reasons that sugar is such an absolute disaster for health and longevity.

Sugar substitutes aren't any better. Many people (including me when I was overweight) turn to artificial sugars to quell their cravings without packing on the pounds. Back then I would have happily performed heart surgery with a Diet Coke in my hand if only I could have found a way to sterilize it! But ironically, although these products are supposed to aid in weight loss, they do just the opposite. That's because products such as sucralose, saccharin, aspartame, and other nonnutritive artificial sweeteners kill your gut buddies and allow the bad bugs to multiply. Believe it or not, a Duke University study[28] showed that a single Splenda packet kills *50 percent* of normal intestinal flora!

It's sad but true: if you eat too much of anything sweet, your gut buddies will starve to death, and your bad bugs will live long and prosper—and multiply. Even fructose, the sugar in fruit, has been shown to be a mitochrondrial poison! There goes the neighborhood.

This may all sound pretty bleak, but rest assured, it is possible to undo the damage these bad guys have wreaked on your inner condo association and get your property back to its youthful, gleaming state. But before we do that, it's time to learn how to keep the bad bugs where they belong.

Chapter 2
Protect and Defend

Hopefully by now I've convinced you of just how significant a role your holobiome plays in both how long and how well you live. Fostering the right mix of microbes in your gut, in your mouth, and even on your skin is essential not only to preventing disease but also to living a long and joyful life. And if you have a plentiful and diverse majority of gut buddies residing in your inner condo, they will keep you young.

This seems simple enough, right? Just starve out the bad bugs while feeding the good ones plenty of the foods they need to thrive, and you and your gut buddies will ride off into the sunset together. And I can stop writing and you can stop reading.

Unfortunately, it's not that simple. Having the right gut buddies in your microbiome is only half of the equation. The second half is making sure they stay on their side of the intestinal border. When pieces of their cell walls called lipopolysaccharides (LPSs, which we will talk more about in a minute) cross the border from your gut to your organs, tissues, lymph, or blood, it doesn't matter if they're gut buddies or bad bugs. Any bacteria, LPSs, or other invaders lurking where they don't belong trigger an immune response that generates widespread inflammation and lays the groundwork for accelerated aging and illness.

As Hippocrates famously and wisely said, "All disease begins in the gut." The good news is that all disease can be stopped there as well.

THE OTHER SIDE OF THE TRACKS

To better understand how inflammation leads to aging—a concept that has come to be known as *inflammaging*[1]—let's first take a closer look at how the gut wall functions. Your intestines are lined with a single layer of mucosal cells (called enterocytes), which are locked tightly together to prevent material from entering or escaping. Though this intestinal layer is only one cell thick, its surface area is equivalent to the size of a tennis court.[2,3,4] The immune cells (specialized white blood cells) positioned along and interspersed within that lining play an immensely important role in maintaining the integrity of the wall—in fact, about 60 percent of all your immune cells are concentrated along the lining of your gut. These immune cells are responsible for deciding what can leave the GI tract and what must stay contained.

Your stomach acids, enzymes, and gut buddies break down the food you eat into individual components: amino acids (from protein), fatty acids (from fat), and sugar molecules (from sugars and starches). Your mucosal cells then literally bite off a single molecule of these digested amino acids, fatty acids, and sugars, pass it through the body of the cell, and release it into your portal vein or lymph system. When all is working well, everything besides these single molecules remains outside the intestinal barrier, where it belongs. There is a lot of truth to Robert Frost's poem that says, "Good fences make good neighbors." If your mucosal cells are lined up tightly side by side, your gut lining functions as a "good fence" that keeps everything except single molecules of digested amino acids, fatty acids, and sugar on the other side. But if your fence gets worn down and becomes rife with microscopic holes, it will allow other compounds to leak through, and your health will begin to suffer. This is the definition of "leaky gut," also known as intestinal permeability, and it is at the heart (or shall I say gut?) of most of the common diseases we associate with aging. In fact, as you will soon see, it's the gradual breakdown of this barrier that accelerates the aging process.

That's because when the wrong molecules or even bacteria cross the border, the immune system kicks into high gear. It's helpful to think of the immune system as the police force of your internal condo. When they learn that someone has broken in, the cops flood the scene and call for reinforcements by releasing inflammatory hormones called cytokines. It's great to have these "cops" around when you really need them. For example, if the "bad guy" that breaches your intestinal lining is truly dangerous, such as a bacterial infection, they can save your life. When you have an injury, inflammation can help you heal. But problems occur when the cops get called over and over for every little thing. The result is chronic inflammation, the ultimate cause of the common diseases of aging, from Alzheimer's to cancer, diabetes, and autoimmune diseases.

To reiterate, aging is so closely linked to inflammation in the body that my fellow longevity researchers have coined the term *inflammaging* (I wish I'd thought of it!) to describe the fact that human aging is characterized by chronic, low-grade inflammation.[5] Though this is a catchy term, paradoxically I have observed that inflammation is actually a *symptom* and not the root cause of what really ages us. Instead, aging is the result of a lack of the right bacterial population in your gut, along with a leaky gut that allows bacteria and other particles to pass through the intestinal border between you and them. When I help my patients heal their gut wall and balance their gut buddies, their inflammation levels (which I can measure based on the amount of cytokines in their blood) decrease dramatically, and their bodies rapidly repair the damage, just like restoring an old house with "good bones."

But what can damage your gut wall in the first place, inciting a constant inflammatory attack by the cops? One such culprit is lectins, which pry apart the tight bonds between the mucosal cells that line your intestinal wall. Hopefully lectins never make it to your gut in the first place, either because you didn't eat them or because the mucus (collectively called mucopolysaccharides, meaning "many mucosal sugars") in your nose and mouth and esophagus bound and trapped them before they could travel down to your gut lining. Those sugars exist for this very purpose; lectins like to bind to sugars. If lectins do make it to the gut lining, the mucosal cells there are the next line of defense, hopefully producing plenty more mucus to bind and trap lectins before they can breach the gut wall.

But for most of us, that protective layer of mucus is in short supply or doesn't exist at all. If your diet is high in lectin-containing foods, that mucus is constantly used up binding those lectins. What's worse, without mucus, the mucosal cells that produce mucus are then open to direct attack by acids, bacteria, and more lectins. The result is less protective mucus.

Unfortunately, without mucus to trap them, lectins bind with receptors (organelles that respond to a signal) along the gut lining and produce a compound called zonulin, which breaks the tight junctions that hold together your border wall, just like that kid in a game of "Red Rover" who broke open the interlocked arms of you and your friends. Imagine this happening on a large scale across your tennis court–sized border, creating spaces between the cells so foreign invaders (including more lectins) can forge through and reach your tissues, lymph nodes, and bloodstream.

Once across the border, these foreign proteins are recognized as foreign by sophisticated bar code scanners called toll-like receptors (TLRs) located on your immune cells, especially your T cells. Off go the air-raid sirens, an all-points bulletin is called out, and the race is on for the cops to apprehend these interlopers! Now imagine this happening every minute of every day, and presto: chronic inflammation!

When lectins poke holes in the fence, they're not the only ones that can make it through—they also clear the way for other invaders, including the bad bugs in your gut. One particularly harmful example, which will be familiar to you if you've read *The Plant Paradox*, is lipopolysaccharides (LPSs), molecules that make up the cell walls of certain bacteria in the microbiome. Though I'm usually not one for cursing, I can't resist calling these guys "little pieces of shit," because that's exactly what they are! LPSs are fragments of bacterial cell walls that your bacteria produce as they divide and die in your intestines. They are produced in the trillions every day, and when your gut border is breached, they travel through your gut wall and into your body. But you don't even need a leaky gut wall for them to gain access into the rest of you, as they can also hitch a ride on specialized saturated fat–carrying molecules called chylomicrons. More on that later.

Now, here's the problem: LPSs are not living bacteria; they are just the outer surface of bacteria. Since your TLRs on your immune cells cannot tell the difference between LPSs and living bacteria, they assume the LPSs are the real thing and call in the troops. After all, the existence of bacteria in your body could mean a potentially deadly bacterial infection. The cops have to be on high alert. So every time an LPS slips past the border, the cops flood the scene, and the result is more and more inflammation.

It gets worse. Lectins have a molecular pattern that is similar to that of LPSs and are foreign proteins, so when they cross the border, the TLRs light up, and the cops come running yet again. My friend and colleague Professor Loren Cordain of Colorado State University (the "father" of the paleo diet) first described this phenomenon as "molecular mimicry": lectins mimic the proteins on many of our important organs, nerves, and joints. The immune system mistakes these parts of our own bodies for foreign invaders and launches an inflammatory response. This "friendly fire" process is the root of all autoimmune disease. It is the result of trespassers getting across the gut wall and latching on to parts of your body while your immune system attacks without asking questions first, and it all arises from a disruption of the complex ecosystem that is our microbiome.

The latest studies conducted on mice prove that this inflammatory response is also a major cause of aging. In 2018, researchers at the Yale School of Medicine correlated a microbe that was present in mice with a lupuslike autoimmune condition that crossed from the gut into the mice's organs. The result was gut wall disintegration and immune cells (which you can think of in this case as mouse cops) in the same organs as the invading bacteria.

Notably, the same bad bugs were found in liver biopsies of human patients with autoimmune diseases, but not in healthy control subjects.[6] In other words, a leaky gut that allows bacteria to cross the border of the gut lining causes autoimmune disease in both mice and humans.

But do not despair. As I recently published in the journal *Circulation*, of the 102 patients with biomarker-proven autoimmune diseases who followed my program, 95 (94 percent) were disease-free, off their medications, and had negative biomarkers after six months.[7] Many of my colleagues and I now believe that this disruption of the gut wall barrier is at the heart of inflammaging, as well as the multitude of diseases that accelerate aging.

You might be thinking that leaky gut is a condition that affects only a few "canaries in the coal mine"—but sadly, this is not the case at all. I'll admit that when people first started talking about leaky gut, I was skeptical. If you had asked me about it fifteen years ago, I would have dismissed it as fanciful thinking for the most part. But there is no denying that it is real and it is ubiquitous. As my treatment program for thousands of patients has shown, a leaky gut (and that includes a "leaky" mouth and nose and skin) is the root cause of aging and disease. As my colleague Dale Bredesen has shown, abnormal bacteria entering the brain from the nose and sinuses can cause Parkinson's disease, while still more studies have implicated bacteria and other microbes as causing atherosclerosis. Other researchers and I now believe that 100 percent of the population has some degree of intestinal permeability.[8]

SPACE INVADERS

Eliminating lectins is an important step in healing your gut and slowing (and reversing) the effects of aging, but lectins aren't the only molecules that pry apart the mucosal layer and cause intestinal permeability. One of the most common causes is, ironically, the use of nonsteroidal anti-inflammatory drugs (NSAIDs) such as ibuprofen, Naprosyn, Aleve, Advil, Celebrex, and Mobic.

These drugs were introduced in the early 1970s as an alternative to aspirin, which was known to damage the stomach lining. However, we now know that NSAIDs damage the mucosal barrier in both the small intestine and the colon. A great deal of research published over the last half century reveals that these apparently "harmless" drugs actually blow gaping holes in the intestinal barrier. As a result, lectins, LPSs, and living bacteria are able to flood into your body. Inundated by these foreign invaders, the cops that make up your immune system do what they do best: attack, producing inflammation and more pain. Having holes in your gut lining is painful in and of itself.

This pain in turn prompts you to down another NSAID, promoting a vicious cycle of pain and inflammation. For years, doctors like me were clueless about the real effects of NSAIDs (although it appears that the pharmaceutical companies knew about them all along). In fact, these drugs were viewed as so dangerous that when they were introduced, they were available only by prescription and only for two weeks. Unfortunately, the available technology limited physicians' ability to fully assess the damage. Because our gastroscopes didn't reach all the way to the small intestine, we had no way to see the damage these drugs were causing. It wasn't until we had microcameras that patients could swallow, giving us 360-degree views of their digestive systems, that we were able to see what was really happening—and by then, NSAIDs were being popped like candy. NSAIDs are now both the number one pharmaceutical seller and the number one cause of inflammation—the very thing they're meant to treat!

Another class of drugs that is disastrous for your gut is proton pump inhibitors (PPIs) and other stomach acid reducers such as Zantac, Prilosec, Nexium, and Protonix. Stomach acid is important and necessary. It kills off most of the bad bugs you swallow before they make it to your gut. Without enough of it, bad bugs—including those that can cause infectious diseases—can take over. This is why people who regularly use acid blockers are three times more likely to get pneumonia than those who don't use them;[9] stomach acid is one of the best defenses against bad bugs getting into you, as one of its main purposes is to kill bacteria. Also, remember that lectins are plant proteins; stomach acid is designed to digest proteins. So by using stomach acid blockers, you inadvertently wipe out one of your major defense mechanisms against lectins!

But back to our gut buddies: most of them hate acid. And some of our most important microbes live in the colon, where there is no oxygen and no acid. What keeps them there is referred to in medicine as "the acid gradient." Stomach acid gradually reduces as food moves along the intestines, where the liver and pancreas add other alkaline digestive enzymes. This transition to a low-acid environment happens where your small bowel meets your colon. But with no stomach acid to keep them in their place, bacteria can easily crawl from their home in the colon into your small intestine, where they don't belong. There they disrupt the gut barrier, causing leaky gut and laying the groundwork for a condition called small intestinal bacterial overgrowth (SIBO). Here these bacteria really wreck havoc by living in a place without much defense against them, thereby laying waste to the absorptive surface of your gut and producing the protein wasting and muscle loss so often seen in the elderly. What's worse, SIBO and irritable bowel syndrome (IBS) are associated with an increased risk of dementia, as shown in a recent Taiwanese study.[10]

Stomach acid is so important to protect your gut barrier that my colleagues at the Medical College of Georgia (where I went to medical school) are starting to use baking soda as a treatment for autoimmune diseases such as rheumatoid arthritis. When patients drink a baking soda solution, it stimulates the production of more stomach acid. Among other things, this additional acid helps to keep the gut bacteria where they belong and prevents inflammation—and therefore helps to reverse autoimmune disease.[11] Moreover, their research shows that baking soda actually sends a signal to the immune cells lining your gut to "chill out" when confronted by foreign proteins.

Proton pump inhibitors do more than neutralize acid and generate inflammation. These medications are aptly named: they paralyze proton pumps, which your mitochondria (hello again, sisters!) need to generate energy. When PPIs were introduced years ago, we were naive enough to think that they worked by paralyzing only certain proton pumps in the stomach lining that make acid and not the proton pumps throughout the rest of the body. Unfortunately, that is not the case. So, every time you swallow that Prilosec OTC to enjoy your corn dog heartburn free, it doesn't stop at the stomach but actually poisons your brain's mitochondria (and the mitochondria throughout the rest of your body as well), making it impossible for them to produce energy.

Sure enough, a long-range study published in 2017 that looked at nearly 16,000 healthy people aged 40 and older found a significant association between cumulative PPI use and the risk of dementia.[12] Meanwhile, a German study from 2016 showed a 44 percent increased risk of dementia among 74,000 people aged 75 and older who used these drugs, compared to those who did not.[13] Other studies have linked the use of PPIs to chronic kidney disease.[14] Not surprisingly, these diseases can all stem from mitochondrial dysfunction. No wonder the FDA issued warnings on their use and package inserts advise not to take these drugs for longer than two weeks. How long have you been on them?

Because you need stomach acid to break down dietary protein into amino acids before they can be absorbed in your gut, people who take these drugs are also likely to be protein malnourished. This is not because they aren't eating enough protein. Rather, it is because they have no stomach acid to digest it into amino acids! Couple that with a leaky gut from SIBO and lectins, and it's no wonder that you hear doctors telling their older patients that they need to eat more protein. But when protein isn't broken down and absorbed, it leads to muscle wasting, called sarcopenia, a health crisis among the aging population. I mean, why live to 100 if you don't have the strength to walk to the kitchen (or worse, the bathroom!)?

Okay, okay, *enough*, you're saying. *What am I supposed to do about my heartburn, Doc?* I get it. I used to have horrible heartburn, or gastroesophageal reflux disease (GERD). I refused to use PPIs, but there was an ever-present supply of Tums and Rolaids on my nightstand and in my suitcase. It's now been more than seventeen years since I've had heartburn, because I don't eat the lectins that were the cause of the heartburn in the first place. Rest assured that the foods you'll eat on the Longevity Paradox program will not cause heartburn. Many hundreds of my patients have been able to throw out their PPIs to protect their gut buddies (and their homes) without suffering any negative consequences. Indeed, several of my patients with the precancerous condition Barrett's esophagus were completely cured by stopping their PPIs and removing lectins from their diet.

AND YOU THOUGHT GLUTEN WAS BAD

Though most lectins, like gluten, are too large to get through the gut wall unless it's already been breached, there is a lectin called wheat germ agglutinin (WGA) that is very small. So even if the gut mucosal barrier has not been compromised, WGA can pass through the walls of the intestine and cause inflammation, notably in the kidneys.[15] But WGA also causes many other problems in the body, particularly because of its ability to mimic insulin.

Bear with me for a moment while I take a quick detour to explain how insulin normally works in the body. It's critically important to understand this science, as insulin resistance and type 2 diabetes have become the norm among the aging population and having diabetes dramatically increases your risk of developing cancer and other diseases of aging.[16] In 2018, the American Diabetes Association reported that more than 84 million Americans had prediabetes and therefore an increased risk of developing type 2 diabetes as they aged, along with an increased risk of heart disease and stroke. (And as I tell my prediabetes patients, diagnosing someone as "prediabetic" is like telling a woman she is a little bit pregnant!) At the same time, a full 25 percent of Americans over the age of 60 had full-blown diabetes.[17] That's one in every four older Americans.

So no matter how old you are, it's important to learn how to avoid diabetes and stop prediabetes in its tracks.

Here's what you need to know. Normally, when the cells in your gut bite off a single molecule of sugar and pass it into the bloodstream, the pancreas responds by secreting insulin, a hormone that regulates the amount of sugar (glucose) in your blood. Insulin works by opening the door to your cells and allowing glucose to enter. The mitochondria in both muscle cells and neurons then "digest" the glucose, using oxygen to create energy. When there is more glucose available than your muscles need (and if you are sitting all day, they don't need much!), insulin attaches to a docking port on the cell membrane of fat cells and flips an enzymatic switch called lipoprotein lipase that tells the fat cell to convert that glucose to fat and store it in the cell. No matter what type of cell, once the insulin has done its job of ushering glucose inside, the insulin separates from the docking port so the cell is ready to receive the next hormonal signal.

This system goes completely awry when WGA is absorbed after you eat whole wheat products. WGA locks onto the insulin hormonal docking port on the cell membrane, but unlike a "normal" hormone, which would dock, release its information, and leave, WGA doesn't let go. Instead, it keeps the switch open. This causes fat cells to continue ushering in more and more sugar to store as fat. In contrast, in muscle cells, real insulin cannot dock at the port because it is blocked by the WGA. This means that your mitochondria can't get the glucose they need to create energy. Without energy, your cells die. Many people assume that muscle wasting is a normal part of aging, but that is not the case. This insulin mimicry is one of the main causes of muscle wasting as we age because muscle cells die when they cannot get glucose for energy. But don't panic; muscles and the rest of you have a backup power supply that we'll get to shortly.

Worst of all, WGA locks onto the docking port of nerve cells, starving them of energy, too. Your brain needs a lot of energy to function. So when sugar is locked out of your neurons, your brain responds angrily, demanding that you eat more food. Of course, your brain cells don't understand that the sugar from any food you eat will just continue to go straight into your fat cells because the insulin docks on all of your other cells are being blocked. No matter how much you eat, your muscle and brain cells will starve and your fat cells will feast. Hardly a recipe for youthful longevity! In addition to fat gain and reduced muscle mass, over time this can cause brain cells and peripheral nerves to die, resulting in dementia, Parkinson's disease, and peripheral neuropathy.[18] Wow, consider that the next time you order whole wheat toast in an effort to live longer!

ONLY THE STRONG SURVIVE

So besides avoiding WGA, how can you maintain your gut wall integrity? You may be surprised to hear that this goes back to the idea of hormesis. As you read in chapter 1, hormesis is the favorable response of an organism to low doses of stress that would be harmful in larger doses. And it actually plays a big role in activating longevity. This is because of how your gut buddies and their sisters, your mitochondria, respond to stress.

Remember the example of the roundworms that lived longer when they were given supplements of a compound that is normally released by gut bacteria? The same idea applies to you. When your gut buddies believe there is an incoming threat, they initiate a response that protects against infections, tumors, and even death. This is why, in one study, mice that were subjected to low levels of radiation throughout their lives lived an average of 30 percent longer than their unexposed siblings.[19] Other experiments using environmental stressors such as heat, cold, lack of nutrients, ultraviolet light, and toxins all came to the same surprising conclusion: at the right dose, these potentially lethal factors can actually promote survival.

Alcohol is another great example of a hormetic stressor (a type of stress that stimulates hormesis).

All studies that look at the effects of alcohol over time show a classic hormetic curve, which simply means that drinking some alcohol is good for health and longevity and drinking a lot of it is bad. One study that followed more than five hundred healthy men for a sixteen-year period found that those who had an average of two small servings of alcohol a day lived longer and had lower rates of heart disease than those who abstained, consumed alcohol rarely, or overindulged.[20]

I first studied the impressive benefits of hormesis when performing heart surgery. Before stopping the heart for a prolonged period of time to repair it, I briefly shut down the flow of blood to the patient's heart. This period of stress alerted the patient's heart cells that trouble was on the way, activating a complex series of events ushered by a compound called heat-shock protein, which caused the heart muscle cells to hunker down and protect themselves until a better time (in this case, better blood flow) arrived. Weak cells in that area that could not survive the stress were either eaten by white blood cells or instructed to die in processes called apoptosis and autophagy, respectively. The result was that only strong cells remained and weak ones were eliminated, enhancing the patient's chance of survival during and after surgery.

In the Longevity Paradox program, you will take advantage of a hormetic stressor called calorie restriction, which significantly extends life span in all creatures studied to date, including our cousins rhesus monkeys. Wait a minute, I can hear you say, you want me to eat less food? No, thank you!

Before you put the book down, hear me out. I've been practicing this method for more than seventeen years now, and if you have ever dined with me, you know that I eat a lot! What I've learned from experiments on thousands of patients and from following the research of my colleagues studying longevity is that there are actually ways of tricking the body into *thinking* that you are fasting or severely restricting calories even when you are eating plenty. You'll read much more on this later. But for now let's look at how calorie restriction promotes longevity.

Self-Devouring Cells

Autophagy is your cells' recycling program, the process by which they get rid of weak or dysfunctional parts of the cell to make it stronger overall. The Latin *autophagy* means "self-devouring": the cell literally eats the pieces that it wants to get rid of. Autophagy is a natural process that is triggered by compounds in certain foods. It also happens when you stress your cells temporarily. When they get the signal that tough times are on the way, your cells develop an "only the strong survive" mentality and toughen up to help you and them endure the impending challenge. When the cells lining your gut go through this process, the gut barrier is actually reinforced so that fewer invaders are able to make it to the other side. The result is less inflammation and disease and a longer, healthier life.

In 2018, the results of a ten-year study in France offered the first conclusive proof that calorie restriction extends the lives of primates—in this case, a gray mouse lemur that shares many physiological similarities with humans. In the study, the life span of the calorie-restricted lemurs increased by almost 50 percent, and best of all, so did their health span. The aged lemurs had the same motor abilities and cognitive performance as much younger animals, without suffering from other common diseases associated with aging such as cancer or diabetes.[21]

How does this work? Well, one of the first things to happen when you restrict calories is that you dramatically decrease bacterial growth and reproduction. Hey, if you give your bugs less to eat, they make fewer offspring. This means fewer LPSs. A second bonus of eating less is that you aren't eating as many foods containing lectins. These two factors dramatically decrease the amount of bacteria, LPSs, and lectins crossing the gut wall, which automatically reduces inflammation. Calorie restriction also improves gut wall function by stimulating autophagy in the gut,[22] thinning the herd of gut bacteria down to the strongest and fittest ones that will work the hardest to maintain your gut wall integrity.

Now pay close attention, because this is where things get really interesting. Remember, the cells lining your gut produce mucus. There is a family of bacteria (called *Akkermansia muciniphila*, which literally means "mucus loving") living in that mucus layer that love to eat the mucus lining. In other words, these gut buddies will never starve because even when you restrict calories, they happily live off that mucus. Now, that sounds like a real problem. You need that mucus to protect you against lectins and other bacteria, and these guys are eating it! But here's the shocker: amazingly, when they eat the mucus, they send a signal to your enterocytes to produce more mucus. So, even though they are eating some of it, the net effect is an increase in mucus.

The more mucus you have, the thicker your wall of protection from invaders, which means that calorie restriction works in part by improving the integrity of the gut wall.[23] As you know, the breakdown of our gut border unleashes mischief and is a main cause of aging. Thus, these mucus-loving gut buddies are one of the keys to your longevity. Indeed, mice that are given supplements of *Akkermansia muciniphila* show a decrease in inflammation and heart disease, because the muciniphila keep the gut barrier strong and impermeable.[24]

Further, having plentiful muciniphila is inversely correlated with obesity, diabetes, and inflammation.[25] When researchers fed muciniphila bacteria to fat mice, the mice quickly lost weight and showed a decrease in blood sugar levels.[26] This suggests that these mucus-loving gut buddies can help prevent type 2 diabetes. And amazingly, it turns out that the popular antidiabetes drug metformin actually works by changing the gut biome.[27] In fact, a study from Columbia University that tracked 459 participants who took the drug revealed that metformin modified their gut microbiota, leading to a higher relative abundance of none other than the muciniphila bacteria![28]

Back to our friend *Akkermansia muciniphila*. There's only one problem: your population of this important gut buddy declines naturally with age.[29] So in the Longevity Paradox program, you are going to give them what they want and, in doing so, ensure that you have plenty of them around for the long haul. As a teaser, we have known for centuries that a certain type of fermented green tea called pu-erh, which is popular in China, has many health benefits. Sure, you know that tea is good for you, but pu-erh tea? Go to the head of the class if you guessed what it does: it promotes the growth of *Akkermansia muciniphila* bacteria![30] Eating and drinking foods like these and periodically practicing (or mimicking) calorie restriction can dramatically increase your *Akkermansia muciniphila* population to strengthen your gut barrier.

Your cells also respond to calorie restriction by becoming more efficient at producing energy. They do this by growing more mitochondria in a process called mitogenesis. Since mitochondria have their own DNA, they can divide and multiply within a cell, without the cell itself having to divide. Simply put, the more mitochondria you have, the more energy you have and the more efficiently your cells function. Think about how fuel economy standards forced carmakers to produce more horsepower with less fuel. Thanks to turbocharging and superchargers, a four-cylinder engine can produce the horsepower of a V-8 on a third of the gas (energy). When your cells think that food is going to be in short supply, they essentially transform themselves into turbocharged engines that are packed with mitochondria, which produce more energy with less food.

STEM CELLS AND LONGEVITY

Many of my fellow longevity researchers view stem cells, undifferentiated cells that have the potential to become pretty much any type of cell when they divide and multiply, as the holy grail of longevity. Indeed, when you recruit stem cells from the body, you can regenerate aging tissues. Seeing this, many doctors have developed stem cell therapies that involve pulling stem cells out of a patient (from his or her fat or bone marrow, both places where stem cells are plentiful), spinning them in a centrifuge, and reinjecting them. Many claim to have experienced wonderful results from these treatments, but I believe you can get similar (or better) results without having to spend thousands of dollars and recover from a procedure. After all, you have plenty of stem cells in your body already. So how can you make them perform the way you want them to?

As we age, our stem cells start to lose their ability to regenerate—unless you activate them by creating a signal that turns on the switch and tells them to do so.[31] Temporarily stressing your cells activates that switch and recruits stem cells from all over your body to regenerate. We know from mouse studies that when mice fast for twenty-four hours, their cells begin to use fat for fuel instead of glucose. This process is called ketosis, a term you are likely familiar with given the popularity of the keto diet. Ketosis creates stress in the body and signals stem cells to regenerate.

In a study conducted by my friend and colleague Dr. Valter Longo at the University of Southern California, participants who underwent a five-day vegan calorie-restricted diet once a month or a few days of water-only fasting successfully shifted their stem cells from a dormant state to a state of self-renewal.[32] Longo also found that the fasting process triggered autophagy in immune cells. As old and damaged immune cells were killed off, stem cells swarmed in and differentiated themselves into healthy new immune cells.

These findings are promising because having healthy immune cells is a pillar of youthful longevity. Your immune cells protect you against everything from cancer to bacteria, so you need them to be as strong and robust as possible as you age. As a general rule, the older you live, the weaker your immune system becomes. This is why the elderly are at an increased risk of dying from common infections such as the flu—their immune system often just isn't up for the fight. On the Longevity

Paradox program, we will give your immune cells what they need to stay healthy and vital as both of you age.

Stem cells are also located in your gut. On that tennis court–sized surface area of your intestines are millions of tiny hairlike projections called microvilli. At the base of each of these microvilli are crypts, and living inside these crypts are bacteria and stem cells.[33] Now, the intestinal surface grows and dies incredibly quickly. The stem cells in your gut constantly repopulate the microvilli with new enterocytes (intestinal cells), as enterocytes, because of the "heavy lifting" involved in nutrient absorption, are constantly being shed into the lumen of your intestines. As I like to tell my patients, think of this process as akin to the lines of soldiers doing battle in the Revolutionary War or Civil War. As the first line of soldiers is shot and wounded and fall, the next line of soldiers behind them steps up to take their fallen comrades' place.

On those stem cells in the gut are docking ports called G-protein receptors—I call them G-spotters for short. These receptors receive chemical signals that cause them to activate and keep up healthy new intestinal wall growth. One of those signals is called R-spondin, which is generated by—you guessed it—your gut buddies![34]

I like to think of these crypts as hardened bunkers

for your condo association. In case of a disaster, the population will survive because of these few key gut buddies that are hidden away. And when your body needs more stem cells to repair your gut wall or other parts of your body that are degenerating, the gut buddies in the crypts will send a signal to activate the stem cells that are in hiding with them. When you restrict calories and the mucin-loving bacteria start chomping away at the gut lining, the gut buddies in the crypts send the signal that reinforcements are needed up at the front line. This causes stem cells to proliferate and migrate up the microvilli to repopulate the gut lining. So at its core, fasting is good for you because it's "bad" for you.

Another signal that activates the stem cells in your gut is vitamin D3. Without adequate levels of this essential vitamin, the stem cells remain inactive, even when your gut lining is being degraded. If the lining degrades too much, it shrinks down to the size of a Ping-Pong table rather than a tennis court. This, in turn, can cause malnutrition.[35] You simply can't absorb as much nutrition through a surface that small no matter what you eat.

In one remarkable new study, severely malnourished children were administered extremely high doses of vitamin D3—200,000 IU—a day in an attempt to help aid nutrient absorption. Happily, the result was that the children all rapidly gained weight despite not consuming any more calories. With a rejuvenated gut wall, they were able to absorb the nutrition in the food they ate.[36] Is it any wonder that we naturally seek the sun? Maybe our gut stem cells and our gut buddies are asking for a little help down there?

There is also a little-known connection between stem cells and telomeres, the endcaps of our DNA that protect our precious genetic material against damage or "fraying" as we grow older. Telomeres have lately become a hot topic in the field of longevity, with many people positing that shortened telomeres are the cause of cognitive decline and accelerated aging. We know that as we age, our telomeres shorten and our chromosomes have less protection against damage—and that damage (aka genetic mutations) can lead to diseases such as cancer and Alzheimer's. But I believe the science is still out as to whether telomere shortening *causes* aging—or vice versa. We do know that telomeres within an organism can have differing lengths and the longest telomeres occur in the colonies of stem cells living in the crypts. In one study in mice, shortening of telomeres in specific crypts paralleled a decline in stem cell functionality.[37]

This suggests that telomere length and stem cell activation go hand in hand. And you already know that stem cell activation is dependent on your gut buddies. So there is a pretty clear link between a happy population of gut buddies and your ability to maintain healthy, long telomeres as you age. Moreover, humans with the highest levels of vitamin D3 in their blood, compared to those with the lowest levels, have the longest telomeres and vice versa.[38] Hmm . . . vitamin D helps stem cells, and vitamin D promotes long telomeres . . . I guess we better make sure to get lots of vitamin D! Don't worry—vitamin D toxicity is extremely rare. In fact, one prestigious study suggested that 40,000 IU a day can be consumed without producing toxicity. That same study found that the average person would need 9,600 IU a day to have an adequate level of vitamin D.[39]

THE CYCLICAL NATURE OF LONGEVITY

Your gut buddies evolved to endure cycles of growth and regression. That's because we, their host organisms, evolved to eat in tune with the seasons—to consume different foods in different quantities at different times of the year. Eating seasonally is more than just a culinary trend: it is embedded in our DNA. In fact, when a large group of researchers from California, London, and Canada worked together to study one of the few remaining hunter-gatherer populations in the world, the Hadza tribe of Tanzania, they found striking differences in their gut microbiomes among seasons, in a cyclical configuration.[40] Specific types of bacteria grew and flourished in their gut based on which foods were available to them. The researchers compared this to people living in a modern urban setting, who had access to any type of food they wanted any time of the year. They had no such variability in their gut buddies among seasons.

The microbiomes of chimpanzees and gorillas, too, fluctuate with seasonal rainfall patterns and diet, changing markedly from the dry summer period, when there are plenty of fruits in season, to the rest of the year, when they eat a more fiber-rich diet of leaves and bark.[41] Indeed, all great apes gain weight only during fruit season, and they don't have 747s transporting fruit from Chile in February. We are meant to live in sync with seasonal shifts. Spring and summer are periods of growth and reproduction (and thus call for a higher energy intake), while fall and winter are times of regression and retrenchment (and thus call for a lower energy intake). But in our modern society, we live in a 365-day growth cycle with abundant calories of all kinds and no natural opportunity to reset. As a result, our population of gut buddies stays relatively the same all year long.

Part of eating for longevity means reestablishing our connection to this all-important cycle. And that means changing up the types of foods that you eat and periodically limiting the amount of calories you eat. A lot of my patients initially fear this part of the program, worrying that they'll be uncomfortably hungry if they go more than a few hours without a snack. (You can thank the junk-food marketing companies and the US Department of Agriculture for the idea that you need to eat every two hours . . . but I digress.) The truth is that humans are really good at fasting, or at least limiting food intake. It's what has allowed us to endure periods of famine or traveling immense distances in search of food. During times of food scarcity, our mitochondria rev up as our entire immune system hunkers down and kills off any cells that are odd, inefficient, dysfunctional, or weak.

It's also during times of scarcity that our fat stores are used. Before we invented domestic refrigerators, only about a hundred years ago, we stored glucose and excess protein as glycogen, a form of sugar we can store in our muscles and liver, and used that sugar for energy until more food was available. If we used up all of our glycogen before we got more food, our mitochondria were able to switch from "digesting" glucose for energy to "digesting" a fat-derived fuel called ketones. This is a critical part of our beautiful human design. But our 365-day growth cycle has caused most of us to lose this ability. We never run out of food and we eat foods in no relationship to the seasons, so our mitochondria don't have a chance to switch from burning sugar to burning fat.

This may not sound like a big deal, but I assure you it is. It takes your mitochondria only half of the effort to utilize fat (from ketones) and turn it into energy as it does with sugar. Your mitochondria prefer fat as a fuel source. But they rarely have the opportunity to burn fat because most of us rarely experience periods of scarcity anymore. Instead of getting to use the type of fuel they want, your mitochondria are busy trying to process all of the sugar that's constantly getting dumped into your cells.

Remember, insulin ushers sugar into muscle cells; if those cells are "full" (think of your muscles saying that they couldn't eat another bite), any excess sugar or protein that you are eating is shunted for storage as fat into fat cells, for use, hopefully, during a future famine. But if you eat yet more sugar and protein before that famine hits (which of course you will), your pancreas will keep releasing insulin, trying to get that sugar out of your bloodstream and into your cells. Over time, this leads to weight gain, type 2 diabetes, a shortened life span, and a greatly reduced health span. Even more distressing, that excess sugar in the bloodstream is the ideal food source for cancer cells, which are able to reproduce easily during this constant growth cycle.

In addition to the seasonal cycle, the day/night cycle is also a key factor in longevity and health span. All animals have a twenty-four-hour circadian rhythm of sleep and wake cycles based on periods of darkness and light; animals that are exposed to artificially shortened or lengthened periods of daylight have shorter life spans. To better understand this phenomenon, we need to look at how and why animals go to sleep. Most of us are well aware of how vital sleep is, but just like seasonal variations, the circadian rhythm of sleep is essential for activating longevity. In a study in 2012, when humans were restricted to five hours of sleep for four nights, they began to develop insulin resistance (prediabetes).[42] Why does this happen?

Humans and many other animals have a pair of neuron clusters within the brain called the suprachiasmatic nucleus (SCN), which receive light input from the retinas of the eyes.[43] Amazingly, when old animals receive SCN transplants from younger animals, their life span increases. But we can use other techniques to promote longevity via the SCN. Diet, caffeine, and resveratrol, found in red wine, all influence the SCN. Calorie restriction may be so effective at promoting longevity by the actions we described above because it helps synchronize the SCN. This further proves that we are meant to spend some time in a calorie-limited state.

Another benefit of cycling between lots of food and not much food is that when calories are restricted, an essential gene called the SIRT1 gene is activated. This gene is also activated during sleep, because you're (hopefully) not eating while you're asleep. And guess what the SIRT1 gene controls? The suprachiasmatic nucleus! When researchers at the Massachusetts Institute of Technology (MIT) blocked the SIRT1 gene activity in young mice, their circadian rhythm control was impaired, and aging was accelerated. Eating automatically shuts off the SIRT1 gene, and a deficiency of this protein interferes with your ability to sleep. Yet when animals were given supplements of the sleep hormone melatonin, their SIRT1 gene protein production was enhanced.[44] And guess how melatonin is made naturally in your body? By your gut buddies, of course. What goes around comes around!

Your body is designed to be a part of a beautifully orchestrated dance between periods of sleep and wakefulness, eating and fasting. By prolonging the period between waking and eating (in other words, extending the "fast" before "breaking" it with our morning meal—break-fast, get it?), we can extend our health span by continuing to activate our all-important survival genes.

STRENGTHENING THE GATE

In addition to melatonin, your gut buddies produce other important hormonal signals that help offset the 365-day growth cycle and help strengthen the gut wall. Butyrate, a short-chain fatty acid, is one example. Only certain gut buddies produce butyrate, which improves mitochondrial function, modulates fat and glucose metabolism in the mitochondria, and has anti-obesity and antidiabetic effects.[45] Butyrate can therefore help counteract some of the negative effects of living in a 365-day growth cycle. Butyrate also protects you from cancer by inhibiting cancer cell growth,[46,47] and there is strong evidence to suggest that it promotes brain health by increasing mitochondrial activity in the brain.[48] For example, when mice with advanced Alzheimer's disease were given butyrate, it significantly enhanced their learning abilities.[49]

And guess what else? Butyrate produces ketones in the liver, which you already know your mitochondria love.[50] So your gut buddies actually do their sisters a favor by producing butyrate for them! The more prebiotics (foods that are beloved by specific beneficial gut buddies) you eat—and you'll be eating plenty on the Longevity Paradox program—the more butyrate your gut buddies will produce.

There are other organic compounds that help protect the gut wall—in particular, polyamines, which are produced by your gut buddies and rely on your gut buddies to transport them into your cells, where they play an important role in cell growth, differentiation, and survival. In addition to protecting the gut wall, polyamines have strong anti-inflammatory properties, promote autophagy, regulate brain function, and have been shown to promote longevity in many animals.[51]

Studies have consistently shown that higher levels of polyamines are linked to increased autophagy and longevity. For example, when researchers in Japan gave mice supplements of polyamine-producing gut bacteria, the mice showed suppressed inflammation, improved longevity, and protection from age-induced memory impairment.[52] In another study, rodents given supplements of polyamines throughout their entire lives had a 25 percent increase in life span. Even rodents that were fed polyamines only late in life experienced a 10 percent life span increase. The researchers credited this to polyamines' role in stimulating cell autophagy, which kills off weak and abnormal cells and strengthens the overall organism[53,54]—and I couldn't agree more.

Sources of Polyamines

Just to get your juices flowing, take a look at some of these good nutritional polyamine sources:

- Shellfish such as squid, oysters, crabs, and scallops
- Fermented foods such as sauerkraut
- Cruciferous vegetables
- Leafy greens
- Mushrooms
- Matcha green tea
- Nuts and seeds, including hazelnuts, walnuts, and pistachios
- Chicken liver
- Aged cheeses
- Lentils

It is also possible to supplement with polyamines, which we will discuss in greater detail later.

In addition to feeding your gut buddies polyamines so that they will protect your gut wall, on the Longevity Paradox program you will be eating plenty of polyphenols, plant compounds that nourish gut buddies and stimulate beneficial processes such as autophagy. The best-known and most powerful polyphenol is resveratrol, which is found in grapes, red wine, and berries and is the reason red wine is protective against heart disease. Resveratrol stimulates autophagy through a different pathway than polyamines do,[55] so it is essential to get plenty of both types of compounds to make sure that your cells recycle themselves as efficiently as possible.

As we conclude this chapter, remember that the integrity of your gut lining is essential to your health span and longevity, and it is under attack at all times. On the Longevity Paradox program you are going to protect it from all angles by:

- Feeding your gut buddies the foods they love, which will help them create compounds that support your mitochondria
- Tricking your body into thinking you're fasting so it will prune your gut buddy population and all of your cells down to the strongest ones

- Thickening your protective gut lining and mucosal lining to keep invaders out

With the hordes stopped at the gate and the residents happy as clams, your inner condo association will become a peaceful community. Ready to get started? Before we begin building that community, let's make sure you're 100 percent clear on what is aging you and what is not. Here's a hint: it's probably not what you think.

Chapter 3
What You Think Is Keeping You Young Is Probably Making You Old

When new patients first arrive in my office, many are suffering from multiple diseases that we associate with "normal" aging. Oftentimes they are perplexed by their illness—they think they are already doing everything right to live a long, healthy life. But commonly the very things they think will keep them young are actually causing them to age more quickly. There are a lot of popular theories of aging out there, and many of them are flat-out wrong. Yet they are deeply ingrained in our culture and in fact may seem completely logical on the surface. I call these the Seven Deadly Myths of Aging, and it's time to debunk them once and for all.

THE SEVEN DEADLY MYTHS OF AGING

MYTH 1: The Mediterranean Diet Promotes Longevity

We can learn a lot about how to age successfully from people who live in the so-called Blue Zones, a term coined by the journalist Dan Buettner to describe five parts of the world where people live the longest. Yet many discussions about what sets these communities apart and what they have in common are rife with half-truths and flat-out myths. So let's take a look at what is really going on in these extraordinary places where people live to the age of 100 at *ten times* the rate of the US population.

These locations are the Ogliastra region on the Italian island of Sardinia; Okinawa, Japan; Loma Linda, California (where I once lived and worked as a professor at Loma Linda University); the Nicoya Peninsula of Costa Rica; and the Greek island of Ikaria. Not included in Buettner's list are a few other locations whose residents are famously long lived: the Kitavans in Papua New Guinea and the residents of Acciaroli, a small town south of Naples, Italy. Many so-called health gurus look at the list of Blue Zones and, seeing that two of the longest-living cultures are found on islands in the Mediterranean, advise their followers to simply follow the Mediterranean diet, which includes grains.

But a closer look at those cultures reveals that cereal grains are actually a negative component of the Mediterranean diet,[1] meaning that these folks live long, healthy lives *despite* eating so many grains, not because of it. In fact, because of their reliance on grains, Italians overall have significantly high rates of arthritis,[2] and Sardinians in particular have a high proportion of autoimmune diseases.[3] Our gut buddies, even in these long-lived communities, still haven't adapted to eating grains, and that includes the new "in" grains such as quinoa and farro. As many of my Peruvian patients tell me, their mothers taught them always to pressure-cook quinoa to remove the toxins; and farro is just wheat, plain and simple, but with a gussied-up name.

Although there is some overlap among the nutrition patterns in these groups, in fact the populations of the Blue Zones have wildly different diets. Let's take a closer look.

- The Seventh-Day Adventists in Loma Linda eat a lot of nuts and soybeans in the form of textured vegetable protein (TVP). This meat substitute is made from high-heat and high-pressure-extruded defatted soy meal—and having lived there and eaten these for many years, I can tell you that TVP can be made into any "mystery meat" you care to name, including a look- and taste-alike for Spam called Wham! You bean lovers out there, take note: TVP is pressure-cooked soy; pressure-cooking destroys lectins. Those clever Adventists! The majority of Seventh-Day Adventists are vegetarians or vegans, yet—spoiler alert—their diet (even the vegans') is made up of 50 percent fat. Apparently a 50 percent fat diet promotes longevity; more on that later.

- The Nicoyan staples include corn tortillas, beans, and rice.

- The Sardinians with extreme longevity live in mountainous areas away from the coast, so they eat little fish, but they have goat cheese and goat meat, eat a bread made from buckwheat and wheat, and consume enormous amounts of olive oil.

- Ikarians also consume a lot of olive oil, as well as herbs such as rosemary, a weed called purslane, which you can see growing along any sidewalk in the United States (the summer annual moss rose or portulaca is the same plant), and regularly drink wine at breakfast!

- The Okinawans eat very little fat (what fat they do eat comes from pork lard) and almost no tofu or rice. And the rice that they do eat is white, not brown. Roughly 85 percent of their diet is made up of the purple sweet potato, a carbohydrate.[4]

- The Kitavans smoke like fiends and eat huge amounts of taro root (a carbohydrate) and coconut (a saturated fat). Yet they are very skinny, have never had a documented heart attack or stroke, and regularly live into their nineties—without medical care, I might add.

- The Acciarolesi eat anchovies and huge amounts of rosemary and olive oil, drink a lot of wine, and eat no bread or pasta but love lentils.

So what do these people all have in common? Surprisingly, it's not what they eat—it's what they *don't* eat. But before I reveal what that is, let's look at the type of carbohydrate that two of these populations (the Kitavans and the Okinawans) consume in extremely large amounts.

Purple sweet potatoes and taro root (as well as plantains and yams) are not regular carbohydrates. They are resistant starches, a subset of starches that behaves differently in your gut than other carbohydrates, such as corn, rice, or wheat, or fruit. Instead of being quickly converted to glucose, which as you know is either burned for energy or stored as fat, resistant starches simply pass through your small intestine mostly intact. These foods are resistant to the enzymes that break down complex starches—hence their name. So eating resistant starches, even in large amounts, generally doesn't raise your blood sugar or insulin levels. This, of course, is key to avoiding type 2 diabetes, obesity, and inflammation as you age. And because they don't cause a blood sugar spike, resistant starches keep you feeling fuller longer than regular sources of starch.[5,6,7]

But the best thing about resistant starches is that your gut buddies absolutely love them! When they eat resistant starch, your gut buddies multiply and produce large amounts of the short-chain fats acetate, propionate, and butyrate. As you read in chapter 2, these create the ideal fuel source for your mitochondria and the enterocytes that line your gut wall. Resistant starches therefore increase your gut buddy population, enhance digestion and nutrient absorption,[8] and foster the growth of the gut buddies that nurture the all-important mucus layer lining your gut.

Is it possible that the Kitavans and Okinawans avoid many of the diseases we associate with normal aging simply because the increased levels of butyrate help their gut linings remain intact? Though we don't have conclusive proof of this, it makes perfect sense to me, especially when you consider the fact that many of the other Blue Zone communities consume foods such as olive oil, purslane, and rosemary that also nourish their gut buddies.

But as I said, I believe that the true secret of these long-lived people lies not in what they do eat but in what they don't eat. And what they don't eat is large amounts of animal protein. With apologies to my friends in the paleo and ketogenic communities, the facts don't lie. Not a single one of the Blue Zone populations consumes significant amounts of animal protein, and I believe that that is their secret to a longer, healthier life. In a randomized eight-week human trial, participants were placed on a 30 percent calorie-restricted diet (in other words, they ate 30 percent fewer calories than they normally would) and then divided into two groups: in one group, 30 percent of their calories came from animal protein, and in the other group, only 15 percent of their calories came from animal protein. Both groups experienced about the same amount of weight loss (fifteen pounds). However, the blood work of the two groups revealed striking differences. The group that ate less animal protein had lower inflammation markers than the group that ate more animal protein, and both total protein consumption (plant and animal based) and animal protein (except for fish protein) consumption were correlated with more inflammation.[9]

Let's review that Blue Zone list again with this knowledge in mind. The Sardinians eat meat only on Sundays and special occasions. The Okinawans eat a plant-based diet that includes only small amounts of pork. Most Seventh-Day Adventists in Loma Linda are vegetarians, and many are vegan. The Nicoyans eat meat only once a week. In Ikaria, a family slaughters one animal per year and then eats the meat in small quantities over the course of several months. And the Kitavans and Acciarolesi eat very little protein, most of it from fish. Humorously, as told by Tracey Lawson in her book *A Year in the Village of Eternity: The Lifestyle of Longevity in Campodimele, Italy,* an elderly man described his relationship with his pig like this: "For a year I feed the pig, then for a year he feeds us!"

Now compare this to the fact that the average American ate *twenty-two pounds* of red meat and poultry in

2018, more than ever before in history.[10] And that doesn't even include other animal products such as eggs, milk, cheese, and so on. Is it any wonder that we're aging more rapidly than ever before as well?

Am I saying that consuming protein from animal products directly ages you? Yes, I am, and that leads me to the next myth.

MYTH 2: Animal Protein Is Essential for Strength and Longevity

I hope the information you just read about the dietary habits in the Blue Zones has begun to convince you that most Americans now consume far more protein, particularly animal protein, than they need. Don't get me wrong—you need adequate protein to power your body and build muscle to avoid muscle wasting as you age. But there is a big difference between the amount of protein you've been led to believe you need and the amount you actually need. This, frankly, is because of commerce, not health.

Animal protein was once the rarest and most expensive type of food and still is in most of the Blue Zones. But in the West, it has become ridiculously inexpensive thanks to government subsidies of the corn, other grains, and soybeans that are fed to industrially farmed animals, poultry, and even fish. The result is that many Western societies vastly overconsume animal protein, leading to higher blood sugar levels, obesity, and a shorter life span.[11,12]

Still not convinced that you should limit your consumption of animal protein? Don't worry; it took me a while to get there, too. As a boy from Nebraska, I grew up eating lots of "healthy" red meat. But my time at Loma Linda University taught me otherwise. My former colleague at Loma Linda Dr. Gary Fraser has conducted a study of the long-lived Seventh-Day Adventists and run a meta-analysis of six other studies.[13,14] His results clearly show that vegan Seventh-Day Adventists who eat no animal products live the longest, followed by vegetarian Seventh-Day Adventists who eat limited amounts of eggs and no dairy products. Vegetarian Seventh-Day Adventists who consume dairy products come next, and Seventh-Day Adventists who occasionally eat chicken or fish bring up the rear in terms of longevity.

Sadly, animal protein is simply not a necessary ingredient for a long health span. As Dr. Fraser has demonstrated, completely avoiding animal protein produced the greatest longevity among an already extremely long-lived people. Further, the risk of developing Alzheimer's disease correlates directly with the amount of meat consumed.[15] For example, when the Japanese made the nutrition transition from the traditional Japanese diet to the Western diet, containing far greater amounts of animal protein, Alzheimer's rates rose from 1 percent in 1985 to 7 percent in 2008.

But why is animal protein so mischievous when it comes to aging? It goes back to the fact that we are meant to thrive in an annual cycle that includes periods of growth and regression. During periods of growth, your cells communicate with one another via a pathway that sends signals for cells to grow and proliferate. This pathway, which is known as the mammalian target of rapamycin (mTOR), helps to regulate cell metabolism, and mTOR itself is a sensor for energy availability within the body. So if mTOR senses that there is plentiful energy in your body, it assumes that you are in a growth cycle. It then activates the production of a growth hormone called insulin-like growth factor 1 (or IGF-1), which sends a signal to your cells to grow, grow, grow. On the other hand, if mTOR senses that there is little energy in the body, it assumes you are in a period of regression, there's little food, and it's time to batten down the hatches, so it limits production of IGF-1.

By measuring your IGF-1 level, we can therefore assess how much mTOR is being stimulated in your body. Most of us nowadays stimulate mTOR constantly because of our plentiful food supply. We always have excess energy in the body for mTOR to sense, so our

IGF-1 levels are routinely high. This leads to disease and rapid aging. When cells are told to grow indiscriminately with no period of regression, it paves the way for cancer cells to proliferate. Your cells also never get the signal to cull the herd and recycle old or dysfunctional cells through autophagy. I've been measuring my patients' IGF-1 levels for years as a marker for aging.[16] And sure enough, both animal and human studies show that the lower your IGF-1 level, the longer you live, and the less chance you have of developing cancer.

But what does this have to do with animal protein? Good question. It turns out that when mTOR is scanning the body for energy availability, it keeps an eye out for certain amino acids more than others. These are the amino acids that are most necessary for growth—methionine, cysteine, and isoleucine—which are prevalent in—you guessed it—animal protein. These amino acids are very deficient in most plant-based proteins. So if you avoid animal protein, you can eat as much plant protein as you want and still trick your body into thinking you're in a regression cycle, so it doesn't stimulate production of IGF-1. This allows you to have your cake and eat it, too—as long as that cake isn't made with animal protein![17]

There is even evidence to suggest that the reason calorie restriction promotes longevity is that it naturally leads to a reduction in animal protein consumption. Researchers at Saint Louis University looked at IGF-1 levels in members of the Calorie Restriction Society, who consume about 20 to 30 percent fewer calories than is considered normal. (If a normal man consumes between 2,000 and 2,500 calories a day, he would consume 1,700 to 2,000 if he were a member of the CRS.) Despite consuming significantly fewer calories, their IGF-1 levels were about the same as those of people who followed a normal diet. The researchers then recruited vegans who were not restricting calories, measured their IGF-1 levels, and found that they were much lower than those of the Calorie Restriction Society folks. Finally, they asked several CRS members to cut their animal protein consumption to zero without changing their total calorie intake, and their IGF-1 levels went down to the same level as the vegans'.[18]

In addition, mouse and rat studies have shown definitively that avoiding the amino acids most prevalent in animal protein extends life span at levels comparable to those that result from calorie restriction.[19] Dr. Valter Longo of the Longevity Institute at the University of Southern California recently received a patent for his ProLon™ protein products, which are based on greatly reducing the amino acids found in animal protein and substituting those found in plants.

Still worried you'll have a protein deficiency if you cut out meat? Don't be. There are plenty of good protein sources—in particular, most nuts and all vegetables—that provide all the nutrition your body needs without the dangerous amino acids it doesn't. Also consider the fact that a 2018 study of men aged 65 and older revealed that higher protein intake had no meaningful health benefit. It did not increase the men's lean body mass, muscle performance, physical function, or any other measure of well-being.[20]

Your protein need is probably much lower than you think. Dr. Longo and I agree that most people require only 0.37 gram of protein per kilogram of body weight.[21] One kilogram equals 2.2 pounds. So a 150-pound man needs about 25 grams of protein a day and a 125-pound woman about 21 grams. You can calculate this for yourself by dividing your weight in pounds by 2.2 to get your weight in kilograms and then multiplying that number by 0.37 to get your required daily grams of protein.

Keep in mind that every day your body recycles about 20 grams of its own protein. Mucus (including that in your gut lining) contains protein, and your enterocytes are primarily protein, and when your enterocytes are shed daily, you digest these proteins and reabsorb them.

Talk about efficient and eco-friendly. As a result, it is nearly impossible to be truly protein deficient as long as your gut buddies are able to digest and help you absorb the protein you do eat. And just a reminder, you use the same metabolism as a bear that fasts for five months and lives off its fat. It emerges from the den every spring very skinny but with its muscle mass intact. If it used its muscles for fuel, it wouldn't be able to hunt come spring. The same applies to you!

And don't worry, you can still take advantage of the benefits of abstaining from meat even if you're not quite ready to give up your steak and eggs. Dr. Longo has shown that doing a five-day modified vegan fast of approximately 900 calories per day once a month produces the same results in terms of IGF-1 reduction and other markers of aging as a whole month of a traditional calorie-restricted diet.[22] You'll be taking advantage of this "cheat" in the Longevity Paradox program to sidestep mTOR and trick your body into thinking that you're hunkering down for the winter in a cave even when you're flourishing in the daylight.

But maybe now you're thinking, isn't growth a good thing? Why are we trying to avoid natural growth hormones such as IGF-1? This is another excellent question, which leads me to the next myth.

MYTH 3: Growth Hormones Promote Youthfulness and Vitality

Think, for a moment, of a poodle. A standard poodle typically lives to be about 10 years old. A miniature poodle or a Yorkshire terrier, on the other hand, can live to be about 20. Now consider the fact that a miniature poodle and a standard poodle have the exact same genes. The miniature version was simply bred over time to become smaller.

Indulge me in a quick sidetrack. Dog breeding actually began in medieval England, where only landed gentry were allowed to own large dogs. In secret, peasants bred large dogs down to a smaller size so they could afford to own dogs that didn't have to eat much and could catch vermin for food. Of course, the smaller the dog, the fewer calories it needs. So is it possible that small dogs live longer than their larger genetic twins simply because they consume fewer calories and are in essence calorie restricted?

It may sound strange, but I think this idea holds a lot of weight (no pun intended). Most of the people who live in the Blue Zones are far shorter than the average height.[23] Think about this, too: it's well established that women have lower rates of coronary heart disease than men and on average live about seven years longer.

Women are also an average of five inches shorter than men. Of course, correlation doesn't prove causality, but consider this: an analysis of 1,700 deceased people found that men and women of the same height actually had the same average life span.[24]

Yet somehow we as a society hold on to the belief that the taller we are, the better. I couldn't disagree more, and I personally find the fact that we're getting taller as a species to be pretty darn scary. Between the end of the nineteenth century and the beginning of the twentieth century, the average heights of both men and women increased by four inches.[25] To put it bluntly, this growth is overrated.

So why are we getting taller? There's no doubt that what and how much we eat plays a critical role. In societies where children consume large amounts of vegetables, they tend to be shorter in stature as adults and begin reproducing later in life than children who eat fewer vegetables. But when those populations switch to more animal and refined-grain sources of food, their growth rates and stature increase. For example, after Western food was introduced in Japan, the population grew significantly taller within just fifteen years.[26] Since the 1960s, people in India and Singapore have also been eating more processed Western food and have grown considerably taller, and their rate of coronary heart disease has shot up in step with their height.

One of the major downsides of constant growth is that it promotes early puberty, particularly menarche (a girl's first menstruation). The early humans' diet, which included cycles of growth and regression, prompted slow growth and late puberty. In 1900, the average age at which a girl began to menstruate was 18. Now it is much younger, with some girls maturing sexually as young as age 8. This very precocious puberty often causes concern for parents, even before they learn that early puberty is linked to a greater risk of breast cancer, heart disease, diabetes, and death from any cause.

The numbers don't lie. Studies of US war veterans, deceased professional baseball players, and French men and women (an odd grouping of folks, but bear with me) all show an inverse relationship between height and longevity.[27,28] Many studies also reveal a connection between height and cancer. In one study, rapid growth during adolescence resulted in an 80 percent increased risk of cancer fifteen years later.[29] Stop right now and read that statistic again—an *80 percent* increased risk of cancer! Want another chilling fact? When I was in medical school in the 1970s, children's cancer wards contained only a handful of beds; now they occupy whole towers or even entire hospitals.

My colleagues conducting another study divided more than 22,000 healthy male doctors in the United States into five categories based on height and followed up with them twelve years later. Even after adjusting for the doctors' ages, the results indicated a positive association between height and the development of cancer.[30] This is terrifying, but it makes sense since high IGF-1 levels, caused by mTOR sensing energy in the body, promotes cell growth. This includes growth of both the cells that help us grow tall *and* the cells that become cancerous. Our shorter ancestors didn't live in a 365-day growth cycle, so IGF-1 wasn't constantly being stimulated the way it is today.

Dr. Longo's studies have also profiled a group of people in Ecuador called the Larons (named after the researcher that originally studied them, Zvi Laron). The Larons, who have absent growth hormone receptors, are unable to make IGF-1. These short adults are free from cancer and diabetes, similar to another group with the same syndrome in Brazil.[31] What's even more intriguing is when you block the IGF-1 receptor in mice, creating "Laron mice," they live 40 percent longer than normal mice. Restrict the calories these mice consume, and they live even longer, yet giving them growth hormone abolishes the longevity effect of calorie restriction.[32,33] This confirms the need to maintain a low level of IGF-1 if cancer-free longevity is your goal; it certainly is mine! Another way to look at this is that if consuming sugars and animal proteins increases your IGF-1 level, then lessening your consumption of them generally (but at least periodically) is the way to go. In other words, periods of promoting regression by lessening food intake in general and sugars and animal proteins in particular not only regulate growth but are important for reducing your metabolic rate.

What's that? You thought that having a high metabolic rate is a good thing? In that case, keep reading.

MYTH 4: A High Metabolic Rate Is a Sign of Good Health

Remember our friend the naked mole rat, who has confounded scientists by never dying of old age? These hairless creatures have an extraordinarily low metabolic rate, something that most of us have probably been told is a bad thing. But the idea that increasing your metabolism will keep you young and trim is an all-out myth. A high metabolic rate is not a sign that you are burning calories more quickly at all; it is a sign that your metabolism is inefficient and working much harder than it should have to in order to burn fuel. Indeed, as detailed by one of my heroes Professor Robert Sapolsky in his book *A Primate's Memoir: A Neurosci-*

entist's Unconventional Life Among the Baboons, alpha baboons with a low metabolic rate get all the chicks and lead a leisurely, low-stress life, while the males with higher metabolic rates have elevated stress hormones and expend far more energy in finding food, and quite frankly, their love life sucks! The females know a loser, guys!

Earlier you read that during times of stress, your cells become more fuel efficient by stimulating the birth of new mitochondria—in essence turbocharging your cellular engines. A high metabolic rate is the exact opposite—like a car that gets only ten miles to the gallon. Your internal condo wants to be as energy efficient as possible. That's why it turbocharges your cells when necessary and recycles everything in your body from the protein in the mucus lining your gut to other usable pieces of dead cells. In fact, the ability of cells to efficiently recycle pieces of dead cells via autophagy and other cellular processes, rather than letting them build up as debris (think landfills in your brain and heart!), is an important recent discovery in longevity research.[34]

In addition to stimulating IGF-1, the main reason consuming animal protein ages you so rapidly is that it requires a lot of energy to metabolize—and if you consume large quantities of meat on a regular basis, your metabolism never gets a chance to slow down.

This is why true carnivores spend much of their day sleeping—to conserve their energy and slow their high metabolic rates.

Just watch your carnivorous dogs and cats for a day, and you'll see what I mean. (You have my permission to take the day off to conduct the experiment.) They sure sleep a lot! Now join me for another day at my nearby San Diego Zoo to watch the giraffes. Nope, not a lot of sleeping going on. The leaf-eating giraffe is the animal equivalent of a high-efficiency, low-polluting engine. Recently, I've seen a lot of patients jump on board the craze of the high-protein "ketogenic" or carnivore diets as a weight loss strategy. Yes, if you're looking to lose weight quickly, consuming a lot of protein will work, but it works in much the same way that a twelve-cylinder sports car works. If your objective is to empty your wallet at the gas station every few miles, it's a great strategy. The sports car will get you up to speed in a heartbeat. But a Prius will get you a lot more heartbeats in the long run!

In the early 1900s, researhers first posited the idea that longevity is inversely related to metabolic rate. They called it the "rate of living." In other words, if you consistently burn energy at a high rate, you will quickly burn out. This is akin to burning the candle at both ends. On the other hand, a low metabolic rate is the "slow and steady" approach to living. It correlates with our naturally cyclical nature of energy production: alternating periods of growth and regression.

I understand that this may be hard to accept at first because it goes against everything you've been taught to believe. Many of my patients initially fear lowering their metabolic rate and along with it their levels of thyroid hormones, the main drivers of your resting metabolic rate. But I have seen that the healthiest centenarians consistently have basal temperatures in the range of 95 to 96 degrees Fahrenheit, not the 98.6 that's considered normal. You use energy to generate heat, among other things, and as you'll see below, running cooler wins. To put it simply, your body wants to conserve energy instead of wasting it revving up your metabolic rate and generating heat.

Some of my colleagues believe that one reason a low metabolic rate promotes longevity is that a slower metabolism reduces the amount of oxidative stress in our cells. When mitochondria use oxygen to manufacture energy, by-products known as reactive oxygen species (ROSs) are produced. These have the potential to damage cells owing to the resultant oxidative stress. One prevailing theory is that oxidative stress is the major cause of aging. Yet when researchers looked at the metabolic rates and markers of oxidative stress in people ranging in age from

20 to over 90, they found no correlation between metabolic rate and level of oxidative stress.[35] There is also no evidence that people in the Blue Zones experience any less oxidative stress than those of us living elsewhere. ROSs may have an aging effect, but they're a very small piece of the puzzle. I suspect that the heat generated by a high metabolic rate is a much larger piece.

The heat generated by a high metabolic rate ages you quickly. This is because when a glucose molecule bonds to an amino acid in a chemical process known as the Maillard reaction, compounds called advanced glycation end products (AGEs) are produced. This reaction is not called AGEing by coincidence! This is one of the strongest chemical bonds known, and heat is required to catalyze the reaction. You can think of it as browning a protein and sugar together, as that is literally what it is. When you char a steak on a grill, the crust that forms is made of AGEs. Have you ever noticed that the higher the heat, the crispier the meat? Well, the exact same thing happens in your brain, in your heart, and even on your skin. The brown "age spots" that show up as you get older and other signs of skin aging are the results of the Maillard reaction.[36]

Many of my patients are delighted to see this hyperpigmentation go away when they begin following the

Longevity Paradox program. In fact, several years ago, a couple in their late seventies returned to Palm Springs in the fall for their yearly "snowbird" migration from Oregon, driving their large RV as usual. During their office visit that fall, the wife told me that her husband had nearly killed them on the road on the drive down. While he was driving, she had noticed that the multiple dark "liver spots" on the backs of his hands holding the steering wheel had disappeared. She exclaimed, "Harry, look at your hands!" With that, he turned the steering wheel sharply to look at his hands and nearly swerved off the road!

We had a good laugh at how witnessing his deaging had nearly ended his newly extended life. But a word of caution: when you witness this Longevity Paradox happening on your own hands (as it did on mine, too), admire it without endangering yourself or others!

In a 365-day growth cycle, glucose, protein, and heat are always present, so you are producing these chemical bonds all the time. The study results are clear that this metabolic state is one of the underlying causes of aging and degenerative diseases.[37] Lowering the heat by strategically lowering your metabolic rate is the best way to reduce these reactions and therefore the rate at which you age. Moreover, since protein, sugar, and heat are required for this action to occur, cutting down on consumption of sugars and proteins is a key component of the Longevity Paradox program.

In fact, your health care provider has probably shown you how fast you are aging with a lab test called hemoglobin A1c (HbA1c). This routine test for diabetes actually measures the crusty bits of sugar and protein attached to your red blood cells. Since red blood cells are recycled about every two months, your HbA1c level gives you an indirect measurement of how fast or slow you are becoming a giant brown spot. In my clinics, anyone who has an HbA1c level less than 5.0 earns a gold star (really). What's yours? If it's above 5.6, you are in big trouble in the longevity race.

There's one more reason that animal protein is incredibly aging, but it's so important that I saved it for its own myth . . .

MYTH 5: It's Important to Get Plenty of Iron As You Age

People develop anemia and become weak and frail as they age because they are deficient in iron, right? After all, Geritol, an iron-rich tonic, was wildly popular in the 1950s and '60s for the treatment of "iron-poor blood." Not so fast. Iron accumulation in the body actually plays a significant role in accelerating the aging process. When researchers in Denmark and Sweden studied millions of blood donors to see if frequently donating blood would lead to dangerously low iron levels, they found that, after adjusting for age and other health conditions, those who donated blood the most frequently lived significantly longer than those who donated blood less frequently.[38] This is because giving blood reduced the amount of iron in their body. (Similarly, one not-so-obvious reason that women live longer than men is that for a good half of your lives, ladies, you part with a significant amount of iron every month.)

In another study that looked at the function of iron, four-day-old roundworms were fed iron and quickly aged to resemble fifteen-day-old worms.[39] At first glance, that may seem insignificant, but a roundworm's entire life span is only about four weeks. So the additional iron actually shortened the worm's life span by approximately one-third! Iron ages us because it interferes with mitochondrial function. As you probably know, iron is a component of hemoglobin, a substance in your red blood cells that transports oxygen throughout the body. And your mitochondria use oxygen to "digest" either glucose or fat molecules to produce energy. So on the surface it would seem that the more iron you have in your blood, the more oxygen can get into the mitochondria and the more energy they can produce. But the opposite appears to be true.

In a 2018 study at the University of Wyoming, researchers looked at the mitochondria of mice and found that those with high iron levels had oxygen *deficits* in the mitochondria. And when they studied mice with Huntington's disease, which causes neurons in the brain to die, their mitochondria, too, had an overaccumulation of iron. Those neurons died due to a lack of mitochondrial function. If your mitochondria cannot access oxygen and produce energy, the cell dies. This provides us with a pathway for understanding other neurological diseases such as Parkinson's disease, Alzheimer's disease, and amyotrophic lateral sclerosis, or ALS (Lou Gehrig's disease).[40]

Indeed, studies on humans show that having increased iron in the blood as a person ages increases his or her risk of developing Alzheimer's disease.[41] And in those without Alzheimer's, brain-imaging technology has revealed a consistent correlation between cognitive dysfunction and iron deposition.[42] There is even a newly recognized form of cell death called ferroptosis that is linked to too much iron in the brain![43] Yet another study on the effects of iron on brain function showed that when Parkinson's patients reduced their iron levels by donating blood, their symptoms were dramatically reduced.[44] Iron is incredibly aging, and it's found in a huge supply, of course, in animal protein.

But here's something interesting: when researchers in Brazil gave rats with high iron levels and signs of memory impairment a single systemic injection of sodium butyrate, their memories improved.[45] Now, remember, if you feed your gut buddies the right foods, they produce butyrate and use it as a signal to tell their sisters, the mitochondria, to rev up energy production. So is iron accumulation in the mitochondria a sign that the line of communication between the sisters has been hijacked, or does it mean that having happier, healthier gut buddies can help protect you from some of the aging effects of iron? Amazingly, it's probably both.

This boy from Omaha really is sad to have to tell you to reduce your beef, pork, and lamb consumption, as well as your consumption of other animal protein sources, but hopefully by now you're convinced of the benefits of limiting your intake of animal protein. So let's move on to some myths about another source of nutrition that is most people's favorite: fat.

MYTH 6: Saturated Fat Should Not Be Demonized

You've seen on the cover of *Time* magazine and you've read in bestselling books that our fear of animal fats is based on fake news and fake studies. Well, I'm eager to debunk the long-held myth in the paleo and keto communities that saturated animal fats such as butter are good for you. The whole idea that these fats were bad for you originated decades ago with the legendary Ancel Keys. In case you're not familiar with him, Keys was a scientist from the University of Minnesota who was tapped by the government to work on issues related to nutrition for soldiers during World War II. Keys invented the K ration, the packaged meals that fed our troops during the war. When President Dwight D. Eisenhower suffered a heart attack in the 1950s, Keys was brought in to consult on the president's diet. Ever since the war, he had been busy studying the impact of diet on health, longevity, and heart disease. He became well known for his Seven Countries Study, which was an analysis of the eating habits of people in seven different countries and their rates of heart disease. The study indicated a correlation between animal fat consumption and heart disease.

Keys presented his findings to the World Health

Organization, and thus the popular belief that fat (particularly saturated fat) was a leading cause of heart disease took root. The McGovern Commission (yes, the presidential candidate George McGovern) used Keys's data to guide the creation of the new government food pyramid that for the first time demonized saturated fats. Thus began the low-fat food craze. Food manufacturers tried to remove the fat from their food products while still making them palatable, so what did they add? You guessed it: sugar! Meanwhile, the Food and Drug Administration and the Department of Agriculture, headed by my fellow Nebraskan and "Cornhusker" Earl Butz, created policies to subsidize the production of corn, wheat, and soybeans and promoted eating whole grain carbohydrates as the cornerstone of a healthy diet. This was the start of the downward spiral we are witnessing in our health and life spans today. Just to refresh your memory, for the first time in history, our life span has now decreased for three straight years.

In recent decades, Keys was roundly criticized and even demonized for allegedly cherry-picking his data. He started out with far more than seven countries in his study, and many people believe that he simply threw out the data that didn't conform to his hypothesis that fat intake causes heart disease. The backlash against Keys led many people to reembrace the saturated fat found in animal products and gave rise to the popularity of paleo and ketogenic eating plans. But I believe that Keys got an unfair shake. More recent analyses of his data show that he did not cherry-pick at all and that even when the data from the rest of the countries are added back in, there is still a clear connection between animal fat consumption and heart disease.

One place where Keys went wrong was in failing to distinguish between saturated animal fat and fat from plants. Follow-up studies have clearly shown that plant fat is negatively associated with heart disease while animal fat is positively correlated.[46] But to his credit, Keys did say that saturated fat (which comes primarily from animals) was worse for longevity than monounsaturated fat (which comes primarily from plants such as olives, nuts, and avocados).

Dear reader, if you've been following closely, you might have already caught what he really missed in his research. Where is animal fat usually found? Yes, it's found in animal protein. Whether it's a juicy steak, pork ribs, salami, or chicken, where there's fat, there's protein, and where there's protein, there's fire! Heat! I have read all of his studies countless times, and he never, to my knowledge, teased out this important connection.

But I came here, unlike Marc Antony, to praise Dr. Keys, not to bury him. Dr. Keys retired to the south of Italy and lived in a village next to Acciaroli, the home of more centenarians per capita than anywhere in the world, where they consume a lot of olive oil. I've had the pleasure of meeting his former housekeeper, who confirmed that the ultimate antifat crusader loved his olive oil! Dr. Keys thrived well into very old age, dying just shy of his 102nd birthday—making him the reigning champion in nutritionists' life span. Sadly, the track record for longevity experts having extended life spans is poor, to say the least.

Herman Tarnower of Scarsdale Diet fame was murdered by his lover at age 69. Nathan Pritikin of the low-fat, plant-based diet that bore his name committed suicide at age 69 after suffering from two forms of leukemia. Robert Atkins died at age 72 from a fall on the ice in New York City; as confirmed by his coauthor, he died obese. The mentor of Valter Longo, Roy Walford, who was the true father of calorie restriction, a member of the Biosphere 2 experiment, and the author of multiple books on longevity, including *Beyond the 120 Year Diet: How to Double Your Vital Years*, died of Lou Gehrig's disease (ALS) at 79. My personal hero, Gayelord Hauser, made it to 89, curing himself along the way of tuberculosis of the hip. The Kellogg brothers, Will and Harvey, both made it to 91, and my friend Jack LaLanne made it to just shy of 97. But Robert

Cameron, the aerial photographer turned diet expert, with his Drinking Man's Diet, a low-carb forerunner of the Atkins Diet of the 1960s and '70s, made it to see the fiftieth-anniversary edition of his book published and passed onward at 98 years old. Not bad for a guy Harvard nutritionists accused of being a "mass murderer"!

In the interest of fair play, still going strong are T. Colin Campbell and Caldwell Esselstyn, both 84 at the time I write this and both advocates of low-fat vegan diets. And my friend in Santa Barbara Patricia Bragg, of apple cider vinegar fame, is 89. But the big winners in the longevity race for nutritionists are Ancel Keys and Luigi Cornaro, the author of *How to Live 100 Years, or Discourses on the Sober Life*, published in 1550, who died at 102. By the way, I realize that writing this book is risky, as some of my critics will be waiting with bated breath for me to kick off soon and prove myself wrong!

So let's look at some of the fat sources that are best for longevity. Not coincidentally, they all come from plants. And Dr. Keys, who was no dummy, chose olives and olive oil. Olive oil's main fat is a monounsaturated fat, oleic acid, but it's not this fat that protects against heart disease, cognitive decline, Alzheimer's disease, and neurological inflammation. It's actually all the polyphenols that are contained in olive oil that make the difference.[47] This is primarily because these plant compounds stimulate autophagy, your cells' recycling program.[48]

Don't forget what sends the signal to your cells to stimulate autophagy—your gut buddies, of course. Our gut buddies love the polyphenols in olive oil, and I try to consume a liter of it a week, as people in many of the Blue Zones do. Nuts are also exceptionally high in monounsaturated and polyunsaturated fat and are remarkably protective against heart disease. Why? Because they, and the prebiotic fiber they contain, are also beloved by your gut buddies! Eating pistachios, walnuts, and almonds (peeled, please) increases your level of butyrate-producing bacteria,[49] with walnuts and pistachios beating almonds in a landslide.[50,51,52]

Some of my friends in the paleo community are also excited about the remarkable effects of butyrate and try to get more of it by eating additional butter (fun fact: butyrate is named after butter). Butter is indeed a tiny source of butyrate, but, sadly, it is not a good idea to eat a lot of dairy products if you want to live a long and healthy life. And that leads me to our seventh and final myth.

MYTH 7: Milk Does a Body Good

Did you notice in our discussion of the Blue Zones earlier that not only do people in all of those cultures eat meat very rarely, but they also consume goat or sheep milk products rather than cow? Call this luck or intuitive wisdom (or flavor preference). Whatever it is, it is clearly one factor that has helped those people live such long and healthy lives.

Here's why. About two thousand years ago, a spontaneous mutation in northern European cows changed the type of protein in their milk from casein A2 to casein A1. During digestion, casein A1 can turn into beta-casomorphin-7, an opioid peptide[53] that attaches to the pancreas's insulin-producing cells and prompts an immune attack (and thus inflammation). This is likely a primary cause of type 1 diabetes.[54] The most common breed of cows worldwide is the Holstein, whose milk contains this problematic protein. Many people notice that milk gives them gastrointestinal problems or causes them to overproduce mucus, which as you know by now is one of our body's major defense mechanisms against other foreign proteins such as lectins, but in most cases, it is the casein A1 protein that's to blame, not the milk (or the milk sugar, lactose) itself.

Furthermore, conventionally raised livestock and their dairy products are laced with antibiotics and Roundup, which will send your gut buddies running for the hills. With some important exceptions that we will discuss later, consumption of dairy products simply is not conducive to a long life and health span. If you choose to eat dairy products, take a note from our friends in the Blue Zones and opt for products made from goat or sheep milk rather than cow. Goats, sheep, and water buffalo were not affected by the mutation, so their milk still contains the healthier casein A2 protein. And good news: most cows in Switzerland, France, and Italy make casein A2, but please realize that most "Swiss cheeses" don't actually come from Switzerland!

And please stay away from plain old milk as a beverage, particularly for your children. Cow's milk is loaded with insulin-like growth factor 1, or IGF-1 (after all, it's designed to make calves grow quickly). Human milk has far lower amounts of IGF-1, as we are designed to be very slow growing. As we mentioned earlier, fast growth is problematic in so many ways. So, no, milk doesn't do a body good.

If you're beginning to panic that I'm about to take away your dairy products, your animal protein, and all of your favorite sources of animal fat, cheer up. You can follow the Longevity Paradox program without becoming a vegan or a vegetarian (although you certainly should head toward that if you want to). But before we get to the program, I want to show you how everything we've discussed so far is connected. Each part of your body—from your heart to your brain to your musculoskeletal system and, yes, even your skin—ages and regenerates for the exact same reasons and via the same mechanisms. And, of course, it all goes back to your gut.

II

Talkin' 'Bout My Regeneration

You now know that much of the decline we attribute to the "normal" aging process is actually the result of the deterioration of your gut wall and your microbiome, as well as the Maillard reaction of heat gluing together sugars and proteins inside you. But this decline is not inevitable. If you treat your gut buddies well, you can get younger as you age. Best of all, your focused efforts on this one area of your body will effect systemic changes. In the chapters that follow, we'll look at how gut health impacts each individual system in the body. When you treat your gut buddies right, you will improve not only your gut health but also your heart health, your brain health, and your joint health, not to mention your weight and your skin. When we're done, you'll have a gleaming new condo that's full of happy, productive inhabitants both inside and out.

Chapter 4
Get Younger from the Inside Out

I often tell my patients that in the end, one of two things is most likely to get them: heart disease or cancer. While other aspects such as mobility and alertness may factor more greatly into health span, it's impossible to enjoy your old age fully if you're dead. So first we are going to tackle the internal organs that tend to age the most quickly.

As a heart surgeon, I have had plenty of opportunities to see firsthand how closely heart health is linked to the health and maintenance of every other part of the body. There is no doubt, in my now vast experience, that if you have heart issues of any kind, you also have major problems throughout your entire inner condo.[1] So, dear reader, we'll start with the heart—why it ages and how you can begin to repair and regenerate this most vital of organs.

HEART DISEASE IS AN AUTOIMMUNE CONDITION

We often think of heart disease as an inevitable part of aging. We're taught that just like the rest of us, the heart gets weaker and weaker as we age, and after a while you'll probably need to go on medication or undergo a few surgical interventions—maybe have your valves replaced and your arteries opened up—until, at last, your heart simply gives out. This is normal, right? After all, when I was in medical school I was taught that heart disease is progressive and that all we can do as physicians is try our best to slow it down. The idea that heart disease is not an inevitability, and that it can actually be reversed without surgery or medication, goes against everything I once believed as a heart surgeon and cardiologist and everything many people and their doctors still believe. But what if I told you that just about everything you've ever been told about heart disease is dead wrong?

After decades of having a front-row view of this organ, I've gotten to know the heart pretty well. And what I've seen with my own two eyes disproves everything I was taught not only about the heart but also about our entire bodies and our health in general. In essence, it all goes back to the immune system—and hence the gut.

Early in my career, many of the adult humans I operated on were smokers who had plaque (blockages) that usually occurred in the first or proximal part of their coronary arteries. My fellow surgeons and I always found that beyond those blockages, these smokers' blood vessels were gorgeous. It made our job relatively easy. We just had to jump past those blockages with a new artery or vein that we could sew onto a nice juicy blood vessel downstream. Plus, most of those smokers, bless their hearts, were skinny and easy to operate on!

As time progressed, more and more of the people with coronary artery disease that I operated on were not smokers. Instead, they had metabolic syndrome, type 2 diabetes, and high insulin levels and for the most part were overweight or obese. And almost all of them, both men and women, had a large amount of abdominal fat.

The degenerative processes we saw in the blood vessels of the nonsmoking patients with metabolic disorders looked completely different from those of the smokers. Instead of beautiful clear blood vessels beyond a blockage, those patients had multiple plaques throughout their blood vessels. Even worse, all of their coronary blood vessels were horribly inflamed. In the places where we were going to place our downstream bypass, the lining of the vessels was full of mushy plaque. In fact, finding a place that looked clear and healthy was the exception rather than the rule. Instead of being enjoyable, doing bypass surgery on these poor folks was downright harrowing.

Later on, I started working in infant heart transplant surgery with my partner, Dr. Leonard Bailey, the founder of this specialty. Dr. Bailey reasoned that if we did heart transplants in babies that were just days old and had no other hope, their immature immune systems might accept foreign hearts as their own, lessening the need for the use of the powerful immunosuppressant drugs we used in adult transplants. (As a quick aside, you see these agents and many of their cousins advertised nightly on television as an easy fix for the autoimmune disease you suffer from. All I will say about that for now is that if you don't have a transplant, why would you take those drugs?)

Well, we were pretty naive back then. As the babies who received the heart transplants grew up, we'd have them come in for routine cardiac catheterizations (where we snake a catheter through a large artery that leads to the heart and can see the heart and the coronary arteries on a screen) to make sure everything was functioning properly. Lo and behold, the blood vessels of those kids looked just like those of my adult diabetic patients! They had plaque all up and down their inflamed coronary arteries.

That was my first clue that all coronary heart disease is immunologic in nature. Even though we thought that we were doing a good job of preventing rejection of the heart muscle itself, the kids' immune systems had recognized that the blood vessels, which were lined with cells from the heart donor, were foreign bodies and had attacked their surfaces. The thickening was a sign of the war going on between the immune system and the foreign proteins lining the blood vessels. Now, that made perfect sense in the transplanted kids because the blood vessels really *were* foreign. Of course their immune systems did not recognize them as friendly and went on the attack. And the thickening was a manifestation of that war, if you will.

But what was causing the diabetic patients' arteries to look like transplant kids' arteries? Were those patients' immune systems attacking their own blood vessels? If so, why? My search for the answer to that question led me to discover much of the information in this book and turned everything I thought I knew about coronary artery disease completely on its head.

I found my next clue when performing heart valve replacements on survivors of rheumatic heart disease, a condition that stems from rheumatic fever and results in inflammation of the heart, blood vessels, and joints. Rheumatic fever develops as a complication of the common childhood illness strep throat, which is caused by an infection of beta-hemolytic streptococcus. If you've ever been infected by this form of streptococcus and you developed rheumatic fever, you probably recovered after some time and things seemed to quiet down. But unbeknownst to you, your immune system formed antibodies to the cell wall of the strep bacterium and is keeping a constant lookout for its presence in your blood.

You can think of this as the police rendering of a suspect on a "wanted" sign that's hanging on every post office and community pin board. Police renderings usually give you a decent idea of what a bad guy looks like, but they're not 100 percent accurate. And this particular rendering happens to look pretty similar to the drawing of the cells composing your heart valve. As I explained earlier, you have toll-like receptors (TLRs: tiny little radars), which are constantly scanning the body for patterns in proteins or foreign substances such as LPSs that match what they are looking for. The cells in your heart valves just happen to contain a pattern that looks almost but not quite like that on the streptococcus cell wall.

So what do you think happens when the cops are on the lookout for that bad guy on the wanted poster and come across someone who looks a lot like him? You guessed it . . . a case of mistaken identity. In survivors of rheumatic fever, this happens day after day, year after year, until slowly but surely the heart valve deteriorates and needs to be replaced. That's where I come in. As I performed those surgeries over and over, I realized that the contents of the dysfunctional heart valves looked suspiciously like the mushy, calcified blood vessels in my overweight, diabetic patients with metabolic syndrome. Cue the lightbulb appearing over my head! In both types of patients, "heart disease" was caused by an immune response, or a police response based on a case of mistaken identity.

Okay, the valves were destroyed in rheumatic fever patients because their cells looked a lot like foreign bacteria, but what was going on with the blood vessels in my diabetics? My next clue, much to my surprise, came from elephants. In the wild, where they eat only the leaves of trees, African elephants have no known coronary artery disease. Due to habitat destruction, however, herds of elephants now graze on grasslands or are fed hay and grains. These animals now have a *50 percent* rate of severe coronary artery disease. Could a simple change in diet really cause such a significant uptick in disease? In a word: yes. Remember, when your microbiome and immune system are confronted with new foreign proteins such as the lectins in single-leafed plants like grasses and grains (which are totally different from the lectins in two-leafed plants), your gut bugs don't have the ability to "eat" these lectins, nor do they have the ability to educate your immune system to tolerate them.

Okay, fine, maybe the elephants' diet was making them sick, but could the same biological phenomenon apply for those of us mammals who weigh several tons less? Well, humans and elephants have more in common than you may think. We both possess a particular sugar molecule that causes lectins to bind to arteries. This lectin-binding sugar, called N-acetylneuraminic acid (Neu5Ac), sits on the lining of blood vessels and the absorptive cells on the gut wall (the enterocytes). Most other mammals have a different sugar molecule, called N-glycolylneuraminic acid (Neu5Gc), on the lining of their gut wall and blood vessel walls. But elephants do not have this molecule and humans lost the ability to make it about 8 million years back, when our gut buddies evolved and we diverged from chimps and gorillas.

Lectins, and particularly grain lectins, bind to Neu5Ac but cannot bind to Neu5Gc. This explains why captive chimps eating a grain-based diet (appropriately named "monkey chow") don't get atherosclerosis or autoimmune disease but the poor grass-eating elephants do. The chimps lack the lectin-binding sugar molecule, but elephants and humans possess it—and it lays the groundwork for heart and autoimmune diseases in spades when we, like those poor elephants, eat grasses and seeds.

TELL PROTEIN TO STOP CALLING 911

The presence of Neu5Ac on our blood vessel and gut lining is yet another reason that consuming large amounts of some animal proteins is so terribly aging. Remember, the animals we eat (cows, pigs, lamb) do not have Neu5Ac on their blood vessel walls. They have Neu5Gc instead. When you eat Neu5Gc, your immune system senses foreign invaders and calls in the defense. But Neu5Gc and Neu5Ac are molecularly very similar. So the "cops" often mistake them for each other. When they're out searching for an invader (Neu5Gc) to attack, they sometimes release their weapons on the unsuspecting Neu5Ac that lines your blood vessels. This is another example of friendly fire that can lead to developing heart disease. But cheer up; fish, shellfish, and poultry all carry our sugar molecule, Neu5AC.

I have seen proof of this immune attack in many of my patients since I've begun using a new blood test that has been clinically validated to predict your chance of having a heart attack or developing angina within the next five years. The test works by assessing biomarkers for damage and repair happening inside your blood vessels. Simply put, if you have an autoimmune disease or lectin sensitivity, this test reveals an autoimmune attack on your blood vessels that is called in by a cytokine called interleukin 16 (IL 16).

If your immune system is the police force of your inner condo, you can think of interleukin 16 as a 911 operator with the assistance of GPS. It alerts the cops to an exact location. This test allows me to see that a high level of interleukin 16 in a patient's blood is a sign that the cops are constantly being called to the blood vessels to find invading Neu5Gc or lectins (or both) and end up attacking the blood vessels themselves.

I've asked many of my patients with markers indicative of impending heart attack issues to remove beef, pork, and lamb from their diets and to limit high-lectin foods—and when I have retested their blood, I've found that their levels of IL 16 have fallen dramatically. In some patients, it has been reduced by half. This means that their chances of developing heart disease within the next five years decreased simply as a result of reducing their intake of certain animal proteins and lectins. Researchers at the University of Paris-Saclay and Loma Linda University had similar results with a much larger sample size. When they looked at the protein consumption of more than 81,000 people over the course of five years and then followed up nine years later to track their rates of heart disease, they found that the people who ate more animal protein were more than one and a half times likely to have died from heart disease over the course of the study than those who consumed protein from vegetarian sources, primarily nuts and seeds.[2]

Pretty compelling, isn't it? I've said it before, but it bears repeating: I'm just a boy from cattle country. I wouldn't have reduced my own animal protein intake so dramatically if the evidence weren't so strong that consuming animal protein creates dangerous inflammation, which is an utter disaster for longevity. In fact, a 2018 study by my colleagues at the Medical College of Georgia at Augusta University found that consuming a single milkshake caused an immune response similar to one provoked by a severe infection.[3]

Furthermore, in 2018, a study confirmed the link between inflammatory bowel disease (IBD, an umbrella term for ulcerative colitis and Crohn's disease, both autoimmune conditions) and heart disease.[4] The three-year study of 22 million patients found that those with IBD were almost twice as likely to have heart attacks as those without IBD. After adjusting for age, race, sex, and even traditional heart disease risk factors, the researchers found that the patients with IBD still had a

23 percent higher chance of having a heart attack than patients without IBD.

These data evidence not only the link between heart disease and autoimmunity but also the fact that all disease—even heart disease—really does begin in the gut. When your gut buddies signal for the cops to attack invaders, the result is autoimmune conditions such as IBD and an increased risk of heart attack. Sure enough, a study of more than six hundred middle-aged women at University Hospitals Cleveland Medical Center found that the more gut bacteria diversity they had, the less stiff their arteries were.[5] When the bad bugs take over, they cause your arteries to become rigid, dramatically increasing your chances of developing heart disease.

There is an old saying in cardiology circles that you are only as young as your arteries are flexible. And there is new evidence that specific chemicals made by your gut bacteria (trimethylamine N-oxide, or TMAO; p-cresyl sulfate, p-cresyl glucuronide; and phenylacetylglutamine, to be precise) actually cause atherosclerosis by stimulating inflammation.[6]

Thankfully, there are polyphenol compounds in red wine and olive oil that remodel and reeducate the gut biome so it stops making these chemicals. All along, we thought that red wine and olive oil protected against heart disease—and they do, but not at all in the ways we suspected. It turns out that they protect you by making changes directly to your gut.

CHOLESTEROL IS AN INNOCENT BYSTANDER

When I explain to my patients that heart disease begins in the gut, the most common response I hear is "But Dr. G., what about cholesterol? Isn't that what causes heart disease?" Well, let's take a look at the real connection (or lack thereof) between cholesterol and heart disease.

Back in the early twentieth century, a Russian scientist named Nikolai Anichkov first put forth the idea that consuming dietary cholesterol caused heart disease. After observing that diseased arterial walls contained up to twenty times more cholesterol than healthy walls, he claimed that consuming cholesterol had led to those changes. Okay, that makes sense. But does this mean that the cholesterol *caused* the heart disease? You astute readers will already know the answer.

One of the main voices questioning Anichkov's theory at the time belonged to none other than our friend Ancel Keys. Keys believed that saturated fat caused heart disease by raising blood cholesterol levels but insisted that dietary cholesterol was irrelevant. It turned out that he was half right. In his studies, Keys found that altering dietary cholesterol levels had relatively minor effects on blood cholesterol. But he did notice that many patients with heart disease had high blood cholesterol levels, and, in an observation that was way ahead of his time, that high cholesterol was associated with insulin resistance and diabetes.[7] He concluded that "all animals, including man, have a high capacity for synthesizing cholesterol" and that consuming even large amounts of it in the diet did not increase one's risk of developing heart disease.

More recent studies have proved that there is no direct connection between cholesterol intake and heart disease. The famous China Study, which looked at the health and dietary habits of people in sixty-five rural Chinese communities back in the 1990s, showed that neither cholesterol consumption nor high blood levels of cholesterol were associated with cardiovascular disease. Instead, it was triglyceride levels in the blood that were positively associated with heart disease.[8] (More on this in a moment.)

Data from the Framingham Heart Study, a highly reputable longitudinal cardiovascular risk study started in 1948 on the residents of a small town in Massachusetts, which is now in its third generation, shows that the omega-3 index (the total amount of omega-3 fatty acids found in the red blood cell membranes) is a far better predictor of heart disease than blood cholesterol levels. In this study, the risk of death in people who had the highest omega-3 indexes (a measurement of the omega-3 fats EPA and DHA in the blood for the previous two months) was a full third less than those who had the lowest omega-3 indexes. (Since omega-3 fatty acids have an anti-inflammatory effect, it makes sense that they can protect you from heart disease.) But when researchers analyzing the data replaced the omega-3 index with total cholesterol levels and looked at the same statistical models, they found that *there was no association between cholesterol levels and heart disease.*[9]

So what is going on here? If cholesterol doesn't cause heart disease, why do people with heart disease have cholesterol in the plaques in their arteries? (And yes, it is in the plaques I see regularly.) I often explain to my patients what I learned from Dr. Michael DeBakey, one of the great fathers of heart surgery, whom I had the pleasure of knowing in his later years. Back in the 1950s, Dr. DeBakey said that cholesterol had nothing to do with heart disease; it was an innocent bystander that got caught up in an inflammatory reaction on the surface of blood vessels.

To understand this, let's take a step back and take a closer look at cholesterol itself. When you eat excess starch or sugars (or protein, which we'll discuss in a moment) anything you don't need for immediate fuel is carried to your liver, where it is transformed into a fat called a triglyceride. Then you need moving vans, which are low-density lipoproteins (LDLs), a type of cholesterol, to carry triglycerides from the liver to cells throughout your body. Once distributed there, they'll be stored as fat or used to make hormones. You actually have at least seven different sizes of these cholesterol moving vans, or LDLs, including four professional moving companies with big vans and strong movers.

This is yet another of the beautifully efficient systems in your body, but problems occur during a 365-day growth cycle when you have so many triglycerides that the moving vans (big, fluffy LDL particles) get filled up. Then you have to call in some less professional guys with pickup trucks (small, dense LDL particles) to strap some more triglycerides onto their backs for transport. These are equivalent to folks heading down the highway with so much stuff strapped to the roof of their station wagon that a mattress ends up flying onto the road and creating a hazard for everyone else. The next time you envision your coronary arteries, think about that mattress blocking the lane! You get the idea.

Now, what about the so-called good cholesterol, high-density lipoprotein (HDL)? They are recycling trucks. They are sent out empty from your liver to go pick up the fat you stored. Ideally, during a season of regression (less food), your body would need to send out lots of recycling trucks to go pick up all the extra stored fat and bring it back to reuse. In other words, during a growth cycle when you're eating a lot of sugars and proteins, your LDL level increases to help carry the triglycerides into storage—but your HDL level drops because you don't need those recycling trucks to go pick it up yet.

Your body is efficient; it will not waste energy producing HDL if you don't need it. But during a cycle of regression, it does. Your triglycerides fall when you're not eating as much, so you don't need the moving vans (LDLs) to take that fat into storage. What you do need are the recycling trucks to go pick it up for reuse.

Very few doctors recognize that the best way to analyze cholesterol data isn't to compare HDL to LDL or HDL to total cholesterol, but rather to look at the ratio of HDL to triglycerides. In fact, in a recent study of more than 68,000 elderly people, no connection was found between LDL level and all-cause mortality.[10] Did you hear that? *None whatsoever.* But a high triglyceride level *is* indicative of health problems. As a guide, your HDL level should be equal to or higher than your triglyceride level, which basically signifies that you're recycling more fat than is being stored. But during our current

365-day growth cycle, the vast majority of people have the exact opposite ratio.

So why do so many people, including your doctor and most of my cardiology colleagues, still believe that high cholesterol levels cause heart disease? Imagine that you're an alien circling our planet observing life down below and reporting back to high command. One observation you could reasonably make is that ambulances cause car accidents. After all, every time you see a car accident, there's an ambulance nearby. That's called an association; it does not prove causality (that one thing caused the other). This may sound foolish, but the same thing actually happens in describing heart disease. When the cops attack your blood vessels, a war is going on there, and cholesterol, like an ambulance, gets caught up in the melee.

This is why cholesterol is incorporated into arterial plaques—not because cholesterol caused the plaque in the first place but because the pickup trucks, the moving trucks, and the recycling trucks all got stuck in the traffic jam following a police attack. In his 1950s wisdom, Dr. DeBakey was right: cholesterol didn't cause the heart disease; it was an innocent bystander that got caught up in the violence.

Many of my colleagues point to the success of statin drugs, which do reduce cholesterol levels and may slightly reduce arterial plaque, as proof that cholesterol causes heart disease. But, if anything, I believe this proves the opposite is true. We used to think that statins treated heart disease by lowering cholesterol levels because as cholesterol levels lowered, so did incidents of myocardial infarctions (MIs). But remember, association doesn't mean causation. We now know that statins actually work by lowering inflammation. With less inflammation, less cholesterol gets caught up in the war zone.

Statins work by blocking the expression of toll-like receptors (TLRs),[11] which, as you'll remember, are scanners that your immune system uses to identify visitors as friends or foes. TLRs look for patterns to identify who's who. In particular, they are on the lookout for LPSs and anything that looks like them. When they find them, they sound the alarm. Statins effectively prevent TLRs from calling the police. The result is less inflammation, less plaque, and less cholesterol caught up in the mix. The reduction of LDL cholesterol is just a side effect of statins; it's not what makes them effective in treating heart disease.

Oh, and one more shout-out to our old friend *Akkermansia muciniphila*, that mucus-loving gut bug. In a study on atherosclerosis in genetically bred mice, *Akkermansia* prevented heart disease even when the mice were fed a Western diet high in animal fat.[12] Talk about protecting the neighborhood!

Many scientists, including me, now believe that when the walls of blood vessels are damaged during an immune attack, cholesterol is used as a patch to repair the damaged wall. There is even some important literature that demonstrates the fact that the more LPSs you have in your bloodstream, the more cholesterol you will have to bind them and prevent them from causing damage.[13] I find this very intriguing, especially in the case of patients with sepsis (a severe infection), whose cholesterol levels suddenly go sky high. Does this mean that cholesterol is a built-in system to absorb invaders?

This theory, which I like, suggests that the reason some people have high cholesterol is that they actually have a leaky gut. When invaders slip past the gut barrier, your body produces more cholesterol in its efforts to sop them up. And sure enough, societies that eat the most nourishing foods for their gut buddies and don't suffer from leaky gut also tend to have low cholesterol levels.

Does this mean that your gut buddies and your gut wall determine your cholesterol levels? It may sound wild to you now, but just ask the Kitavans, the famously long-lived people who eat enormous amounts of coconut oil, which you would think would lead to high cholesterol levels. But the rest of their diet is made up of taro root, which nourishes and protects their gut. And indeed, the Kitavans have extremely low cholesterol levels.[14]

TRIGLYCERIDES ARE THE REAL ENEMY

I tell my patients that I don't worry about their cholesterol levels, but I do care very much about their triglycerides. Consuming any sugar or simple starch will raise triglycerides, and yes, that certainly includes fruit! Fructose, the main sugar in fruit, is actually a toxin that can directly injure cells and disrupt mitochondrial function.[15,16] To cope with it, your body sends most of it straight to your liver, where it is converted into triglycerides and uric acid. The rest of it (about 30 percent) heads to your kidneys, where it acts as a direct toxin to your filtering system.[17] So why do we still think of fruit as a health food? It is healthy if it is eaten as it was meant to be—during the summer, when it is fresh and in season and when we are in a natural growth cycle. When we ate fruit only during the summer, we could handle it in small doses because we then had nine months to detox before the next growth cycle. Eating fruit year-round is terribly aging!

Grains also raise triglyceride levels. Most people have

heard of foie gras, the decadent, extremely fatty duck or goose liver. Foie gras is produced by force-feeding huge amounts of whole grains to those unfortunate animals. They make so many triglycerides in their livers that all the moving vans in the world can't transport it away fast enough. So it builds up, and voilà—they get fatty liver! If you've been diagnosed with fatty liver or nonalcoholic steatohepatitis (NASH), it's likely that you developed the condition from eating "healthy" whole grain goodness and washing it down with fructose-laden juices and fruits.

"Well, okay, then," you might say, "I'll just go on a high-protein diet and avoid all those sugars and carbs." Not so fast. Remember, too much protein is not your friend. And in fact, eating protein also raises triglyceride levels. Believe it or not, you have no real storage system for protein. Your body needs protein for maintenance of cell membranes and intracellular structures and when you are actively building muscle, but it converts any excess protein into sugar, because you do have a storage system for sugar. This process of converting protein into sugar is called gluconeogenesis. And if you make too much sugar, bingo!—you simply convert that into fat or triglycerides. This is why many people who turn to high-protein diets struggle with high triglyceride levels and insulin resistance. The real trick is to avoid high triglycerides by avoiding the consumption of excess animal protein and simple starches.

Contrary to everything I was taught in medical school, heart disease is not caused by high cholesterol levels, as Dr. DeBakey observed so long ago. It is the result of an immune attack on the blood vessels, ultimately caused by problems in the gut.

It All Goes Back to the Gut

It's not just the heart; your gut buddies control all of your internal organs and capacities. For example, most people believe that alcohol causes cirrhosis of the liver, but in fact you can bathe the liver in alcohol all day long and it will never develop cirrhosis. What alcohol in excess does cause is a leaky gut by directly damaging the gut wall, which in turn allows bad bugs and LPSs to enter the portal vein, which delivers them directly to the liver. The cops, called Kupffer cells (no, there won't be a test later), are just waiting for these troublemakers to arrive in the portal triads in the liver, and the battle ensues. When I see elevated liver enzyme levels in a patient's blood work, I know it is a sign of the liver soldiers that have died or been injured in battle.[18] Likewise, the scar tissue that signifies cirrhosis is really an end-stage sign of this inflammation.

Similarly, obese men with fatty liver disease have elevated levels of zonulin in their blood.[19] Remember, zonulin breaks the junctions between the cells lining your gut. So fatty liver disease is a result of a breached gut wall allowing invaders access into your body. New research shows an even stronger connection between fatty liver disease and the gut biome. Specific bad bugs increase susceptibility to fatty liver disease by stimulating inflammation, making you more likely to develop cirrhosis and even liver cancer, while the right population of gut buddies can protect you from inflammation and reduce the severity of disease.[20]

If your gut wall is breached, invaders can stimulate dangerous inflammation at any vascular interface. I treat a handful of patients with pulmonary fibrosis, a lung disease for which there is allegedly no cure. Pulmonary fibrosis is an inflammatory attack on the blood vessels of the lung. When we fix those patients' guts, they experience tremendous turnarounds. One woman who first came to see me while hooked up to an oxygen tank was later able to fly to Europe to go on a hiking trip without supplemental oxygen!

Even hearing loss, another unpleasant "normal" side effect of aging, can be halted by addressing gut health. When researchers at Brigham and Women's Hospital examined the relationship between diet and the risk of developing hearing loss, they found that women who ate plenty of gut buddy–nourishing olive oil, vegetables, nuts, and fish lowered their risk of developing moderate to severe hearing loss by 30 percent.[21]

It *all* goes back to the gut. When you stop eating for the 1 percenters and eat for your gut buddies, they will start taking care of you!

THE CYCLICAL NATURE OF CANCER

By now you understand that the diseases of aging stem from the same root causes: damage to the gut lining and bad bugs taking over your internal condo association. But you might think there is a notable exception to this rule: cancer.

Well, I've got news for you: there are no exceptions to this rule. In fact, new research confirms that our microbiome plays an essential role in determining whether or not we'll develop cancer and, if we do, how we will respond to treatment. When researchers at the

University of Pennsylvania observed that a particular form of cancer treatment was proving ineffective, they administered a dose of antibiotics that eradicated a specific strain of microbes in those patients' intestines (in other words, not a broad-spectrum antibiotic but a targeted one that spared the good microbes) and then reinstituted the cancer treatment. It's probably not surprising to you by now to learn that once the bad gut microbes were eliminated, the treatment became more efficient in killing the cancer cells.[22] Similar results have been achieved in animal studies: wiping out bad bugs with antibiotics leaves cancer-ridden mice with fewer and smaller tumors and reduces metastasis to the liver.[23]

We also know that patients with pancreatic cancer have remarkably similar microbiomes. It is so consistent that it's been called "the pancreatic cancer microbiome."[24] Even more intriguing, these patients have particular strains of gut bacteria present in large amounts in their cancerous pancreases—there are even more of them in the pancreas than in the gut itself. After the bacteria are killed off, targeted immunotherapies become more effective at treating the cancer. This poses the question: What are those bacteria doing in the pancreas in the first place? Is the pancreatic cancer a symptom of bad bugs slipping past a leaky gut and instigating an immune attack on the pancreas? That's not what they taught me in medical school, but it certainly doesn't seem outside the realm of possibility.

Again it goes back to the 365-days-a-year growth cycle in which we currently live that is so rich with opportunity for cancer cell growth. When your microbiome and your cells' energy sensor, mTOR, send the signal that food is plentiful and your cells should prioritize growth, there is no opportunity for your body to take stock and prune any cells that look odd or dysfunctional. This is a deadly problem—literally—because we have abnormal cells all the time. It is normal to have abnormal cells. What is *not* normal is constantly encouraging those abnormal cells to grow by feeding them an endless abundance of energy (aka food). You absolutely must give your body a chance to reset and prune those abnormal cells by regularly but temporarily restricting your energy intake.

Ironically, it's *restricting* energy that allows your mitochondria to more effectively produce energy for your healthy, noncancerous cells. Right now, your mitochondria are constantly working to produce energy from the molecules of food sent to them by their gut buddy sisters. But it's very difficult for them to keep up. Remember, when you eat sugar or protein, your pancreas releases insulin to usher sugar into cells so it can be processed by your mitochondria. But if each cell's receiving dock is already full, insulin has to put the extra sugar somewhere—and it ends up being stored as fat for later use.

This is a great system, but if you keep eating sugar, protein, or WGA (which, as you read earlier, binds to those loading docks), your pancreas has to keep producing more and more insulin to convert the sugar to fat. This is the root cause of insulin resistance, but it also leads to excess glucose molecules that are free for the taking. And guess who wants to take this sugar and use it to grow? Cancer cells. And guess what stimulates them to grow? Insulin! Insulin is another growth hormone; it's Miracle Grow to cancer cells.

Many health enthusiasts try to "cheat" this system by cutting back on their sugar consumption and encouraging their mitochondria to utilize stored fat instead. As you read earlier, this process (called ketosis) is actually a more efficient means of energy production for your mitochondria. So in essence it is a good idea, and I'm always in favor of a good cheat. But here's the rub: your mitochondria cannot process fat directly from your fat cells. Instead, you need an enzyme called hormone-sensitive lipase (HSL) to turn your stored fat into this usable form of fat called a ketone.

HSL is sensitive (hence the name) to insulin. It works

only when your insulin levels are low. When your insulin levels are chronically high, insulin blocks HSL from getting fat from your fat storage depots. And guess what? If you are fasting or reducing your carbohydrate consumption when insulin resistant, your mitochondria sputter to a stop! Water, water everywhere, and not a drop to drink! This is why so many people who try a low-carb diet crash with the low-carb "flu."

So when are your insulin levels low? When you are not eating any sugar or protein! This is when your body allows you to finally dip into its fat storage and produce ketones to "feed" your mitochondria. But since sugar and protein raise insulin levels, a high-protein diet frequently prevents your body from making ketones. This is a big problem with many mainstream "ketogenic" diets that aren't really ketogenic at all because they promote excessive protein consumption, especially in the form of animal protein. This, unfortunately, is another recipe for cancer cell growth.

CANCER AND IMMUNITY

Recall from earlier in this chapter that you have a sugar molecule called Neu5Ac lining your blood vessels and your gut wall, while many of the animals we eat have a nearly molecularly identical sugar molecule called

Neu5Gc that stimulates an autoimmune attack against the lining of our blood vessels—including those in our heart. While this immune response leads to heart disease, it also contributes to cancer growth. As this immune attack is happening, it produces a hormone called vascular endothelial growth factor (VGEF), which attracts blood vessel growth to cancer cells. Then the very same tumor cells also use Neu5Gc to shield them from the cops so they can go unnoticed.

Studies have shown that human tumors contain large amounts of Neu5Gc, even though our bodies cannot manufacture it—which clearly links the consumption of animal protein to the development of cancerous tumors. A recent study at the University of Leeds offered further proof of this link: after tracking more than 32,000 women over the course of seventeen years, researchers noted a significant increase in the risk of colon cancer among women who ate red meat compared to those who did not.[25]

Mitochondria in cancer cells cannot utilize ketones to produce energy the way your healthy cells can.[26] Remember, your body produces ketones when insulin levels are low and sugar and protein are in short supply—during times of regression—and cancer cells thrive on sugar. Cancer cells are also unwilling to utilize sugar molecules to produce energy the way regular cells do when insulin levels are high. Strangely, the mitochondria in cancer cells create energy only through the extremely inefficient system of sugar fermentation. As a result, the average cancer cell needs up to *eighteen times* as much sugar to grow and divide as normal cells.[27] It's therefore very easy to starve cancer cells to death. They cannot grow and thrive without lots and lots of sugar.

We have known that cancer cells thrive on sugar since the 1920s, when a German physician named Otto Warburg discovered that cancer cells have this unique—and therefore vulnerable—type of energy metabolism, which won him the Nobel Prize. But we are now seeing for the first time that another type of cell has the same metabolism—and therefore vulnerability—and that is none other than immune cells, the cops that patrol your body looking for invaders and often attack the wrong suspects. Like cancer cells, immune cells cannot produce energy (and cause inflammation) when insulin levels are low, during times of regression,[28] yet they thrive during a 365-day growth cycle.

Is this proof of a connection between cancer and autoimmune disease, or does it merely show that both illnesses are underpinned by the 365-day growth cycle that is currently leading our health spans downhill? I'm not certain, but what I do know is that when you limit sugar and animal protein consumption and trick your body into thinking it is in a period of regression with the Longevity Paradox program, you reduce your risks of both cancer and autoimmune disease.

Meanwhile, the standard Western diet promotes cancer growth at every turn. Cancer cells love the fructose in fruit, which as you read earlier we are meant to eat only in limited quantities at certain times of the year. Researchers at Duke University have demonstrated that colorectal cancer cells capitalize on the high levels of fructose often found in the liver and promote liver metastases.[29] And a diet high in animal fat inhibits our natural defense mechanisms against cancer cells. Epithelial cells lining the surfaces of organs have the ability to sense potentially malignant cells and remove them from their midst. But in one study, when mice were fed a high-fat diet that made them obese, this defense mechanism was suppressed and the incidence of cancer increased.[30]

So once again, it's time to start taking care of the 99 percenters so that you can prevent cancer and heart disease and start aging in reverse. Here are some good ways to do it.

The Best Foods for Fighting Cancer

What can you eat to starve cancer cells without starving yourself? In addition to the Longevity Paradox program you'll read about in detail later, there are certain foods that have specific cancer-fighting properties. They include the following.

Exogenous Ketones

You know that your body converts stored fat into ketones when your insulin levels and sugar and protein intakes are low, but you can also consume already-made ketogenic foods. Several plant fats contain ketogenic fats. For example, medium-chain triglycerides (found in MCT oil) can be nearly completely converted to ketones, which are the ideal fuel source for your mitochondria. Solid coconut oil (meaning it is solid below about 70 degrees) contains roughly 65 percent MCTs, making it another good source for creating ketones, along with red palm oil, which comes from the date or fruit of the palm and is loaded with vitamin E's tocopherols and tocotrienols and is made up of about 50 percent MCTs. But be wary, palm oil is not red palm oil, and palm oil production is associated with deforestation.

Butyrate, which, as you recall, your gut buddies produce when you treat them right, is a short-chain fatty acid present in small quantities in butter and is also a source of making ketones. But because of the casein A1 in most American dairy products, goat butter, buffalo butter, or ghee (clarified butter, which does not contain protein) are better sources of this ketone precursor than regular or even raw or grass-fed cow milk butter. And remember, sadly, that there isn't much butyrate in butter.

But no matter how many ketogenic fats you consume, exogenous ketones are best used as an energy source when you are transitioning from a diet that is high in sugar, protein, and fat (our Western diet) over to the Longevity Paradox program. Consuming lots of extra ketones in the form of fats isn't going to do you much good if you're still eating burgers and bagels. Remember, most of us walk around daily with high insulin levels that block our ability to convert "love handles" and belly fat into ketones. These dietary ketogenic fat sources can prevent you from crashing during the transition, but you don't need them long term.

Nuts

Nuts (particularly tree nuts) have amazing anticancer properties. In one recent study, researchers at Yale University looked at the rates of death and cancer recurrence in patients with stage 3 colon cancer. The patients who ate two or more servings of nuts a week had a 42 percent reduced rate of cancer recurrence and a 57 percent reduced rate of death.[31] That's right—cancer patients who ate nuts only twice a week cut their risk of death by more than half. That's more effective than most common chemotherapeutic cancer treatments! Notably, there was no reduction in cancer recurrence or death in patients who ate peanuts—not surprising to me (or you by now) because peanuts are a type of legume that is laden with lectins and not nuts at all. In fact, in animal studies, the peanut lectin promotes colon cancer.[32]

In another study, mice that were fed walnuts had less than half as many tumors in their colons than mice that were not fed nuts.[33] To find out why, researchers looked at the mice's fecal samples and examined the bacteria living in their digestive tracts. They found that the gut microbiomes of the mice that had eaten walnuts were similar to one another and favored bacterial communities that protected against colon cancer. In other words, the mice's gut buddies thrived and multiplied on a diet of walnuts and in return took good care of their hosts.

But it's not just colon cancer. In yet another study, this one conducted by the National Institutes of Health in Bethesda, Maryland, where I was a research fellow for two years, people who ate the highest quantities of nuts had a 26 percent lower chance of developing lung cancer than people who ate very few nuts. Amazingly, the benefits were even better for smokers! People who smoked regularly and ate large amounts of nuts had a

39 percent reduced rate of lung cancer compared to smokers who ate very few nuts.[34] This suggests that the nuts actually protected those people from the negative effects of smoking. A systematic review showed that nut consumption is associated with a lower risk not only of cancer but also of death from any cause[35]—and not just by a little bit, either. In one study, women who ate large amounts of nuts reduced their risk of death from any cause by *half.*[36]

Why are nuts so effective at protecting against cancer? Remember, nuts are extremely low in methionine, the amino acid that mTOR is looking for to detect energy availability. You can think of the presence of methionine as a sign that you are in a growth cycle. So eating nuts, which are low in methionine, sends a signal that you are in a period of regression, which helps you fight cancer in all the ways we've discussed. In addition, your butyrate-producing gut buddies love nuts. And you now know that your mitochondria can use butyrate as a source of ketogenic fat! So nuts are really the perfect cancer-fighting food: they starve cancer cells while supercharging your gut buddies and their sisters, your mitochondria.

On the Longevity Paradox program you'll be eating plenty of healthy, cancer-fighting nuts, including:

- Walnuts
- Macadamia nuts
- Pistachios
- Pine nuts
- Hazelnuts
- Chestnuts

Whether we're talking about avoiding heart disease, pulmonary fibrosis, hearing loss, or cancer, the health of your gut influences the health of your entire body. Moreover, the same factors that cause these common diseases of "aging" also cause the other symptoms of aging that we consider "normal": cognitive decline, loss of muscle, joint pain, and rapidly aging skin. In fact, I have seen as a heart surgeon that there is a clear link between heart disease and arthritis. That leads me to the next area of regeneration: the muscles, joints, and bones that you are going to learn how to use to dance your way into old age.

Chapter 5
Dance Your Way into Old Age

I first became aware of the connection between arthritis and heart disease back when I was performing heart surgeries at Loma Linda University. Within five years of undergoing a stent placement or coronary bypass, half of my patients would be back in the operating room for a hip or knee replacement. Amazingly, the opposite was also true: half of the people who came in for joint replacements would be back within five years for a stent or coronary bypass! I started to wonder if the same process that produced arthritis could also be causing coronary heart disease.

There were other clues, too. Because the heart is located right in front of the spinal column, when I looked at a patient's coronary angiogram (basically an X-ray of the heart arteries), I could see all of the vertebral bodies lined up behind the heart on the scan. Early in my career, I noticed that all of my patients with coronary artery disease—no matter how old or young they were—had pretty impressive levels of arthritis in their spines. I thought that was interesting, but my colleagues seemed to attribute it to the fact that the majority of these patients were old. It was just a coincidence that they had both heart disease and arthritis in their spines.

But when I started to see people in their thirties and forties with both heart disease and arthritis in their spines, I knew there was something else going on. I'll never forget one patient—I'll call her Angela—who was in her early forties and had a devastating case of coronary artery disease. When I operated on Angela, I found that some of her blood vessels were so calcified that I couldn't even put bypass grafts on them. Angela also had horrible spinal arthritis. But after her heart surgery, she became a devoted patient and began following my Longevity Paradox program to the letter.

Ten years later, Angela was admitted to our hospital with a terrible case of food poisoning with abdominal and chest pain. Because she had a history of heart disease, the cardiologist on call performed an angiogram to make sure her heart was doing okay. It showed that her heart was actually doing better than okay; one of her blood vessels that had previously been so calcified and so completely closed that I couldn't operate on it had opened up on its own! Angela now had blood flow where she had had none ten years earlier. Even more striking, the angiogram showed that the extensive arthritis in her spine was completely gone.

Impossible, you say? This is the type of turnaround I see all the time in my patients who follow this protocol to the letter. And it's the same turnaround I experienced myself! As I mentioned earlier, I used to have such bad arthritis that I had to wear braces on my knees when I ran. After following my own protocol, that arthritis is history. When you make your gut buddies happy, they will renovate—and that means regenerate—every part of their home.

WHAT REALLY CAUSES WEAR AND TEAR

For years, we believed that arthritis was caused simply by "wear and tear": the older you get, the more you use your joints, and eventually they just get worn out. But your joints don't come with a "use by" date. The most recent research confirms that arthritis is caused not by overuse but rather by bad bugs in the gut creating inflammation. It's that inflammation that "wears and tears" your joints, not aging itself. For instance, when mice with arthritis are given supplements of beneficial bacteria, their systemic inflammation decreases and the breakdown in cartilage in their poor arthritic mouse knees slows down.[1]

It's actually very simple: the combination of bad bugs and a leaky gut releases lectins and LPSs into your body. Lectins bind onto the sugar molecules on your joint surfaces called scialic acid and act like splinters, prompting an immune attack and inflammation that leads to arthritis (and pretty much every other problem that we associate with aging).[2] Picture the red swollen area under your skin when you get a splinter; now imagine that going on inside your joints. Get the picture? Those LPSs make their way into your joints as well and incite the same response.[3] Remember, your cops view LPSs as real bacteria and attack. Amazingly, when we use a needle to draw fluid from an arthritic joint, we find LPSs in the fluid!

So does the wear-and-tear theory relate to what's really happening? In a strange way, it does. Bear with me while I get a little science nerdy here. The cartilage that lines your joints is constantly being remade by groups of cells called chondrocytes and chondroblasts. (You have similar cells that do this for your bones.) During the war in your joints, cartilage is destroyed and regrown, but this occurs unevenly, resulting in literal peaks and valleys of cartilage. The result is a sandpaper-like lining of your joints. No wonder the surgeon tells you that your joint is “bone on bone.”

Is all lost at this point? Not in the least. Your gut buddies can come to the rescue at any time. They can reinforce the gut barrier, prevent lectin and LPS intrusion, and enable us to heal. Those chondrocytes are still there in your “bone on bone” joints and can grow a new articular surface. This recently happened to a patient of mine named Jerry, who was 67 and planned to have his right knee replaced. He came to me for other reasons including diabetes, high blood pressure, and—you guessed it—heart disease. His orthopedic surgeon needed cardiac clearance for the operation. Jerry passed the stress test that I performed but was intrigued by the findings on his blood tests. He had six months before the surgery and agreed to follow the Longevity Paradox program while he waited.

The last time I saw Jerry, he was twenty pounds lighter, no longer diabetic, and no longer taking high blood pressure meds. Great news, indeed! But as we were celebrating, the talk turned to his upcoming knee replacement. “Oh, that,” he said. “I canceled it. Why get an operation when my knee doesn’t hurt anymore?” With that, he hopped off the exam table and skipped around the room to prove his point.

Like Jerry, when my arthritic patients come to see

me, they are rarely suffering from only one disease of aging because all diseases of aging stem from the same root cause. In fact, when they first come in, even my "healthiest" patients are on an average of seven medications that are actually making the root problem worse! Many arthritic patients take over-the-counter nonsteroidal anti-inflammatory drugs (NSAIDs) to relieve the pain and inflammation. The problem, as you already know, is that these drugs blow holes in the gut barrier, making it far more vulnerable to invaders. This all too often becomes a vicious cycle of more pain and inflammation, more drugs, more damage to the gut barrier, more pain and inflammation, and so on and so on until their bodies are completely wracked with disease.

To break this cycle, you need to heal your gut wall and nourish your gut buddies. This will quell inflammation and enable you to stop taking NSAIDs or other pain medications. One of the critical steps in this process is to remove foods containing WGA from your diet. As you recall, WGA is small enough to slip past the gut barrier even if it is intact. WGA is found in the wheat bran, so it is in all whole wheat and whole grain products, including pasta, bread, and crackers, as well as bulgur (cracked wheat), rye, barley, and brown rice.[4]

You may have previously been told that all of these foods are healthy, but I am asking you to start thinking differently for the sake of your long and healthy life. I have seen countless patients grow sicker and sicker the "healthier" they ate, including yours truly. Yet when they adopt my Longevity Paradox program, their bodies begin to rejuvenate. Yes, it is wonderful that we have developed artificial joints that you can turn to when yours give out, just as I am grateful for the surgical interventions that have sustained the hearts of many of my patients. But there is another way, one that will enable you to avoid these interventions entirely as you skip through your long and joyful old age.

KEEP YOUR BONES AND MUSCLES STRONG, HEALTHY, AND HUNGRY

It's not just arthritis that leaves so many elders with limited or painful mobility. Years of chronic inflammation also cause our bones to deteriorate. The result is osteopenia (invisible bone loss) and osteoporosis (a condition in which bones become weak and brittle). This is a health crisis among the elderly. The National Osteoporosis Foundation estimates that 54 million Americans have osteoporosis and that approximately one in two women and one in four men aged 50 and older will break a bone due to osteoporosis.

But humans aren't the only animals that are suffering

from osteoporosis in increasing numbers, and that's because, sadly, we're not the only ones eating a highly inflammatory diet that kills off our gut buddies and allows bad bugs to flourish.[5] In experimental studies, LPSs have produced osteoporosis in mice. Chickens in factory farms that are fed genetically modified corn develop osteoporosis at alarming rates. Ninety percent of commercially raised birds have a detectable gait abnormality due to bone defects, and more than 10 percent die prematurely due to lameness (inability to walk). Think about that the next time you eat a "healthy" chicken breast. That chicken likely had bone defects due to her diet,[6] and when you consume the meat you are also consuming the inflammatory feed that will contribute to your own bone deterioration.[7] On the other hand, postmenopausal women who eat a diet high in gut buddy–nourishing nuts, vegetables, and olive oil have higher bone density along with lower risk of heart disease, diabetes, and cancer.[8]

But perhaps the greatest risk to our musculoskeletal systems as we age is invisible muscle loss. This often happens over time even if we are not aware of it. If you compare side-by-side CAT scans of patients' thighs, first when they were teenagers and then later when they are 40 years old, the scans will look like a cross section of a large round steak because that's what a round steak is—the cow's thigh. In the earlier scans, they show a large amount of muscles with the bone in the middle. Looking at the same thigh years later, the volume is the same, but the composition is quite different. On average, a 40-year-old's thighs have *half* the muscle mass of the teenager's. So how does the overall thigh have the same volume? The other half of the muscle mass has been replaced by fat. That's what we call marbling on a good steak, but unless you are grooming yourself to be devoured as a tasty meal, marbling is not a desirable quality. And where did that cow's marbling come from? Its corn-, wheat-, and soybean-based diet. Where does it come from in the human subjects? The same place.

This means that even if you're one of the lucky people who has managed to stay the same exact size as you've gotten older, it's entirely possible that you've still lost a good amount of muscle mass and gained a lot of fat. This is another paradox, and I see it firsthand all the time when my patients start losing weight and realize they're all flab! They were fooled into thinking they had the same body composition as they did when they were younger because they were the same "size."

Let's take a moment to look at how muscles grow. It helps to think of your muscles as the customer base for insulin, a door-to-door salesman selling a popular product, sugar. When you eat sugar (or protein, which is converted into sugar), insulin goes around to your muscles, knocking on doors and asking "Hey, is anybody hungry?" If the muscles are hungry, they say, "Yes," and they "eat" the sugar. In this case, insulin has an easy job and can knock on plenty of doors throughout one workday. But there are a couple of things that can make the job more difficult for our friendly salesman. The first is if your muscles aren't hungry and they send him away before he can make the sale. The second is if WGA (remember, WGA can get through your gut wall even without a leaky gut) is in the body mimicking insulin.

When either of these things happens, insulin reports back to headquarters that it needs support from a bigger sales force. Your body quickly produces more insulin to help make the sale. But if your muscles still aren't hungry and/or WGA is blocking insulin receptors on your muscle cells, your muscles aren't going to buy that sugar no matter how many salespeople pound on their doors. What's worse, your sales force is blocked by WGA, so sugar can't get in to feed your muscles what they need and they literally waste away. Finally, insulin gives up so it can finish its workday. But first it has to do something with all of that excess product. So it converts the extra sugar into fat by turning on lipoprotein lipase (yeah, I know, I promised no tests), hoping that one day your muscles will be hungrier and it'll finally be able to make the sale.

When this process unfolds consistently over months, years, and decades, the cumulative effect on your body is more fat storage and loss of muscle mass. And voilà—we have CAT scans clearly showing more fat and less muscle on the same-sized thigh. Remember, the second scans were taken when the patients were 40. Imagine what their scans will look like in twenty more years if they don't do something. In fact, I see it all the time in older patients when they first visit me. So many of them suffer from sarcopenia, a severe loss of muscle mass. This is a twofold issue, one of insulin resistance and one stemming from the gut.

Remember that tennis court–sized surface area you have in your gut to absorb protein? Well, after decades of lectins and NSAIDs and bad microbes attacking it, it's now the surface area of a Ping-Pong table. And remember that acid you need in your stomach but can no longer produce because you've been taking proton pump inhibitors (PPIs)? No acid, no protein digestion, and now your Ping-Pong table has literally nothing to absorb. There goes your muscle.

So what can you do to make your muscles hungrier? You've probably already guessed that the answer is exercise—specifically, strength-training exercises that grow muscle mass. When you exercise your muscles, they get hungry and start screaming for food. This makes it easy for insulin to sell them sugar instead of storing it as fat. Your insulin level decreases because the salespeople don't need as much backup, and your fat mass is reduced. The more your muscle mass increases over time, the larger the customer base grows for the insulin salesman. He wants all that business for himself, so he reports to headquarters that he's got this. And you reap the rewards with heightened insulin sensitivity, larger muscle mass, and less fat. But remember this caveat: eating lectins and WGA and popping acid reducers can undo much of the benefit conferred by strength training.

This coincides with another myth about aging: that we age because as we get older we have fewer mitochondria creating energy for our cells. Many of my colleagues believe this is the case because they observe fewer mitochondria in the elderly. But just as reduced muscle mass is a cause of aging rather than a result, so is a reduction in mitochondria.

Your mitochondria have their own DNA (part of the 99 percent) that divides when cell division takes place separately from the division of your DNA (the 1 percent) in the cell's nucleus. As you read earlier, when your gut buddies send their sisters a signal, your mitochondria divide, even if the cell they support isn't dividing. The result is that you have additional mitochondria to extract more and more energy out of what they expect will be less and less available food. Your mitochondria also divide when you are building muscle and need energy to power them.

So we see that two types of stress (calorie restriction and exercise) cause you to turbocharge your cells with more mitochondria. But as they get older, most people don't restrict calories, use intermittent fasting, or do much strength training. The result is less muscle mass and fewer mitochondria for most older people—but this is not inevitable. Slightly stressed cells and hungry muscles will lead to more mitochondria, lower insulin levels, more muscle mass, and overall better health for many years to come.

No matter how old you are, it's never too late to reap the benefits of exercise, but those who start exercising when they're young do have a head start. One study showed that women who exercised at least three times a week as teenagers experienced significantly less height loss (which really means bone loss) after menopause than their peers who did not exercise.[9] And what pulls on bones, keeping them strong as you age? You got it, muscles! Stronger muscles, stronger bones. The clear message is to start exercising now, no matter how old you are.

USE IT OR LOSE IT

As we get older, it's all too easy to convince ourselves that it's okay to exercise less and allow a sedentary lifestyle to creep in. This is what our society deems normal, but the truth is that we have it completely backward. A sedentary lifestyle is what *causes* us to age! People who continue to exercise well into their ripe old age live longer and stay healthier than those who stop moving and allow their muscles to waste away.[10]

My great-grandmother lived actively until the month before her 100th birthday, and until her death she lived in a bedroom on the third floor of her house. Every day, she walked up and down those three flights of stairs multiple times. As a young boy, I thought she was nuts, but now I see that she was actually incredibly wise. Looking back, she reminds me of those centenarians in the Blue Zones, which happen to be mostly in hilly areas. The residents of the Blue Zones walk up and down those hills long into their old age, maintaining muscle mass and dexterity for decades longer than most Americans do.

Both of these examples speak to the importance of working against gravity when exercising. This stresses, and therefore strengthens, more muscles. Hiking, walking up and down hills and stairs, and doing squats and push-ups are all examples of exercises that force you to work against gravity.

If you're worried that you're too out of shape to hike uphill, have no fear. When I was in France a few years ago for a conference, I saw a fascinating study that had been performed in Switzerland by a group of exercise physiologists who were curious about the effects of vigorous climbing on muscle mass development. (Luckily, there were plenty of mountains in Switzerland to test their thesis on.) They split participants in their study into two groups: one group hiked to the top of a steep hill, where a cable car was waiting to bring them back down. The second group rode the cable car up and hiked down. Everyone wanted to volunteer for the second group, right? The study organizers' premise was that the uphill hikers would see greater benefits, but they were wrong. Both groups had the exact same results, while the group that hiked uphill perceived that they'd made a greater effort. When you walk downhill, you're still working against gravity, and your muscles are stressed from constantly having to put on the brakes.

Like my great-grandmother, the people in the Blue

Zones stay active because they lack many modern conveniences. I'm old enough to remember what life was like before we had electric garage door openers, snowblowers, or even a remote control for the TV. (Although, because I practice what I preach, I like to think I look too young to remember such a time!) We had to get out of the car to open the garage door and use our muscles to shovel the snow. As a boy, I had to get up and walk across the room to change the TV channel! And yes, I did walk to and from school (in a foot of snow, in pouring rain, for miles and miles, blah, blah, blah). These may seem like minor things, but they add up to make a big difference in the way we live and the way we age. It's no coincidence that most of the world's longest-lived people lack modern conveniences. They have no choice but to use their muscles as they age, and as a result they age more slowly.

Yet in the West we expect to grow infirm, so we set up our lives to avoid having to use our muscles even when we still can. This becomes a self-fulfilling prophecy. I've seen so many still-nimble folks buy a one-story home or move their bedroom down to the first floor to avoid having to climb the stairs in anticipation of getting older and not being able to climb the stairs. And guess what happens when these people get out of the habit of climbing stairs? They grow infirm, just as they predicted!

The idea that there is a certain age at which we have to grow old and infirm is completely crazy to me. And it's simply not true. Just look at the older people in Australia who were asked to participate in a "ballet for seniors" class and ended up with better flexibility, improved posture, more energy, and an overall increased sense of well-being.[11] Or ask Robert Marchand, the 107-year-old French amateur cyclist. Researchers have studied his physiology and seen that he is indeed getting fitter as he ages.[12]

That's not the only thing that caught my eye when I first read about Mr. Marchand. He is five feet tall and very wiry, which is consistent with everything you've read so far about how disastrous a 365-day growth cycle is for successful aging and that the small shall inherit the earth in their ripe old age. Also of note, he did not exercise regularly for most of his life. It was only after he retired that he began cycling—the opposite of most people, who begin slowing down as soon as they hit retirement age. (They call Palm Springs, where I live, "God's waiting room" for a reason!) Years later, at 107, Mr. Marchand takes no medications and has the aerobic capacity of a healthy 50-year-old.

It is truly never too late to start building muscle and enhancing your health span and longevity. This doesn't mean that you have to spend hours at the gym. I'll share an exercise plan in chapter 10 that is simple, quick, and accessible for anyone of any age at any fitness level. But first let's take a closer look at what exercise does for the body in addition to the important job of making your muscles hungry.

Exercise is another perfect example of hormesis—limited stress you put on your body to make yourself stronger. Like other examples of hormetic stressors such as calorie restriction, exercise stimulates autophagy, the recycling of old, worn-out cellular components, and a similar process called the unfolded protein response (UPR). In the case of UPR, the cell degrades dysfunctional (misfolded) proteins, restoring the health of the cell.

While exercise stimulates both autophagy and UPR, there is evidence that the benefits of autophagy are stronger the earlier you start building muscle. In one study, a group of young men in their late twenties and a group of older men in their sixties—all of whom had not previously exercised—took up a resistance training protocol. Both groups saw an increase in UPR for forty-eight hours after each resistance training session,

but only the group of younger men also saw an increase in autophagy for forty-eight hours after each session.[13] It's never too late to see benefits of exercise, but the earlier you can get started, the better. As model and actress Agyness Deyn is fond of saying, "If you want to stay young, then you better get started early!"

This effect on autophagy and UPR explains why exercise has been shown to minimize the risk of cancer. When left to their own devices, worn-out cellular components and misfolded proteins can cause faulty information in cells, prompting cancerous changes. Rejuvenating the cells through autophagy and UPR helps keep you young and cancer free. It also allows your cells to heal after a trauma, even a heart attack. In a study on mice, exercise was actually shown to produce new heart muscle cells. This was true of healthy mice as well as those that had previously suffered heart damage. Indeed, one reason exercise is so beneficial is that it increases the heart's ability to regenerate.[14]

Regular exercise also dramatically reduces your risk of Alzheimer's disease. A study from 2018[15] showed that women who were physically fit at middle age were a whopping 90 percent less likely to develop Alzheimer's disease even decades later. The few fit women who participated in the study and did eventually develop

Alzheimer's did so an average of eleven years later than women who did not exercise, at the age of 90 compared to 79.

Now listen up, my female readers. As my good friend Maria Shriver and I both know, Alzheimer's disproportionally affects women, and the cure is prevention, not a long-sought-after but not-yet-discovered drug. Imagine that you read a headline saying that taking a "drug" would prevent 90 percent of all Alzheimer's disease if the treatment is started early. How much would you pay for it? Well, that drug is a combination of exercise and, as you'll soon learn, simple choices in food. Another study examined the effects of exercise on patients with early-stage Alzheimer's and found that it improved memory performance and even reduced atrophy of the hippocampus, the memory centers of the brain.[16] We also know that exercise that uses the legs in particular stimulates brain cells, keeping you alert and healthy long into old age.[17]

Remember "Michelle"? I have no doubt that walking her Pomeranian (in her high heels!) multiple times a day helped her stay sharp well into her ripe old age. Meanwhile, "brain training" apps that claim to help you improve your brain actually do nothing for working memory or IQ.[18] So skip the games and go out for a walk instead.

Exercise also has a powerful effect on the immune system. After heavy exercise, antioxidant defense systems are amplified throughout the body.[19] This means that exercise causes you to produce more of the enzymes that support cellular and mitochondrial function. Exercise also helps prevent heart disease, which many of my colleagues attribute to the fact that it reduces inflammation.[20] But you can probably already guess that I attribute it to (drumroll, please) its effects on your gut biome.

That's right—exercise changes the microbiome. Your gut buddies like it when you exercise, so when you do it, they freshen up their house for you.[21] Mice that exercise have significantly higher amounts of a specific type of good gut bacteria (called Firmicutes) than sedentary mice that are fed the same diet,[22] and studies on humans show that exercise increases gut bacteria richness and diversity.[23] Perhaps most intriguing, one study showed that rats that exercised produced more butyrate than rats that did not, all else being equal.[24]

Remember, butyrate protects your gut lining so it is better at keeping out invaders while simultaneously giving your mitochondria their preferred food source. Does this mean that exercise really reduces your risk of cancer, arthritis, and heart disease, all by bolstering your gut wall and your microbiome? I believe the answer is yes. As you've likely heard, exercise also boosts feelings of happiness due to a release of endorphins, the feel-good hormones.[25] Don't forget what creates those hormonal signals for you when you take care of them—your gut buddies, of course! A recent report from the Harvard Medical School stated that for some adults struggling with depression, regular exercise works as well as prescription antidepressants.[26]

Now, you couch potatoes (wait a minute, you're not still eating potatoes, are you?), don't despair. While the research is clear that exercise is an essential component of your longevity plan, even short bouts of the right kind of exercise can dramatically improve both your life and health span. In fact, the plan you'll find in chapter 10 is actually far better for you than spending hours at the gym. This is yet another paradox: a little bit of exercise is essential, but too much of it has the exact opposite effect.

CHRONIC CARDIO = CHRONIC STRESS

Like everything else in life, there is such a thing as too much when it comes to exercise, especially cardiovascular exercise such as running. Our ancestors ran only when they needed to avoid becoming someone else's meal, but somehow we have gotten the idea from faulty calorie math and a misunderstanding of how metabolism actually works that we need to run for several miles at a time or spin, step, or aerobicize for hours on end, all in the name of being healthy. The calories-in/calories-out rule died a horrible death years ago with the discovery of the microbiome and the calories they eat. But the Internet and social media won't let the notion go. The result is that you blame yourself for your unused gym membership.[27]

The long-lived people in the Blue Zones would laugh their heads off at that idea. Back in the 1970s, at the beginning of the jogging craze, researchers interviewed the Kalahari Bushmen in Africa, who are legendary long-distance walkers. They walk between twenty and thirty miles a day during hunting season. When asked about the possibility of running twenty-six miles, they thought such a proposition was ridiculous. They explained that it wouldn't be worth the effort to run after an animal for twenty-six miles because they'd probably burn more calories than the animal contained. And if an animal was chasing *you*, it would catch up with you long before mile twenty-six. In fact, evolutionary biology proves that you will adjust your caloric burn regardless (or in spite of) your exercise program.[28]

Short bursts of speed were useful to early humans when we needed to catch a wounded animal or sprint to the nearest tree to avoid being gored by a wild boar. Other than occasional sprints, however, slow and steady has always won the race. Just look at the body composition of sprinters who compete in the hundred-yard dash—they are muscle-bound, with the exact body type that lends itself to extreme longevity. As I've written previously, chronic marathoners often look as though they are suffering from cancer, and their immune systems are suffering as well.

Besides sprinting, our ancestors' main forms of exercise were walking as they looked for food, and then carrying what they found back to a central location. Replicating these types of movement will give you the greatest longevity benefits. Blue Zone residents are all hikers and walkers, not long-distance runners. And running marathons impairs your immune system. A paper appearing in *Sports Medicine* in 2007 confirmed this sad news.[29] I'll stick to hiking and jogging my four dogs for three miles a day, thank you, instead of running my former 10K and half marathons.

There is plenty of evidence that acute endurance exercise (such as marathon running, which causes you to lose muscle mass dramatically) has a disastrous effect on longevity. Pretty much every research paper that's ever been published on long-distance running shows that it causes myocardial fibrosis, progressive scarring of the heart muscle.[30] It actually damages the heart by killing heart cells, particularly in the right ventricle. I see this in my patients who have been or currently are long-distance runners all the time. The more you run, the more scarring you produce, and this eventually leads to arrhythmias or even congestive heart failure. Though temporary stress has positive effects, long-distance running places too much stress on your heart for too long.

In fact, the exact benefits of moderate exercise are actually *negated* by intense exercise. Exhaustive exercise, exercising to perceived exhaustion, causes oxidative stress by creating free radicals (which as you read earlier are highly reactive uncharged molecules that cause aging), while moderate exercise stimulates an antioxidant effect that protects your body from the oxidative stress caused by those free radicals.[31,32] And here's the kicker: endurance exercise such as running long distances leads to gut permeability.[33]

Bear with me for a quick sidetrack here. I'm old enough to remember not being allowed to swim for an hour after I ate because I'd allegedly develop cramps and drown. My mother actually set a timer for an hour, and I'd sit there eagerly watching the minutes tick by until I was allowed to swim. The digestion process requires huge amounts of blood flow, so after you eat, your blood is diverted to your gut instead of flowing to your muscles and brain. The theory was that if you swam, you'd get cramps from lactic acid buildup (as a result of lack of blood flow) and drown. This has largely been written off as an old wives' tale, but there is actually some truth to it. Blood *does* flow to your gut after you eat. During long-distance runs, the opposite happens: so much blood flows to your muscles and away from your gut that you actually experience ischemia (inadequate blood flow) in the gut. This causes gut permeability, and lectins, LPSs, and bacteria pour into your body. This is precisely why your immune system tanks for about two weeks after an endurance run. And as any runner knows, your digestion gets messed up by long-distance running—a clear sign that your gut is not happy!

I realize that die-hard runners are reluctant to part with their favorite pastime. I used to love running, too, and my wife, Penny, was a passionate marathon runner, qualifying for and finishing the 100th running of the Boston Marathon, until she took a long, hard look at the data. At that point, she could no longer deny the negative effects of her hobby, and she hung up her running shoes for good. She believes (and I wholeheartedly agree) that it was more than worth it to shorten her marathon-running career in order to lengthen her life. And get this: while running, she had osteopenia (loss of bone density) and was constantly getting colds. Now those signs and symptoms of overexercising are a distant memory.

But what good is a long life if you can't remember it? This leads us to the next chapter, where we'll look at some simple techniques for improving your brain health so that you can stay sharp as a tack long into your retirement. Surprise, surprise, even this aspect of your longevity relies on your gut buddies to keep you young.

Chapter 6
Remember Your Old Age

As you get older, things start to slip. You misplace your car keys, grasp for words, and forget the name of your longtime neighbor. Your brain feels foggy, and you're just not as sharp as you used to be. That's just the way life goes, right?

Wrong. Though we think of these symptoms as a normal part of aging, nothing about them is normal. From those seemingly innocuous "senior moments" to more serious neurological conditions such as Parkinson's disease, dementia, and Alzheimer's disease, all cognitive decline stems from the same root cause: neuro-inflammation. And where does inflammation start? In the gut.

But the gut is also where you can put an end to inflammation so that you will stay smart and alert for the rest of your long life. Recent research reveals that you do not need to experience cognitive decline as you age—period, end of story. You have the ability to birth new neurons (in a process called neurogenesis) at any time, meaning there is no expiration date on your ability to learn new skills or grow your cognitive abilities.

In 2018, researchers at Columbia University and the New York State Psychiatric Institute looked at the brain scans of healthy people ranging from 14 to 79 years old, the latter well past the age that most of us consider our prime learning years. They found that the oldest participants in the study had the same amount of the raw materials needed to create new brain cells, or neurons, as the youngest.[1] So no matter how old you are, you can make just as many new brain cells as a teenager! What's more, older adults can improve their capacity to learn new skills and retain memory, particularly their language skills, with increased physical fitness.[2] This means that the fitter you are, the fewer of those "tip of the tongue" moments you'll experience when you struggle to think of the right word or name.

These results offer a lot of hope that you can remain as sharp as ever well into your old age—as long as your gut buddies are motivated to keep you that way.

THE BRAIN IN YOUR HEAD IS ACTUALLY YOUR SECOND BRAIN

There is now so much evidence of the direct connection between gut microbes and the brain that many of my colleagues have begun referring to the gut as the "second brain." I disagree—not with the direct relationship between the two but rather with the gut being relegated to second place. The gut actually controls the brain in your head, which you might want to start thinking of as your second brain.

You already know that your gut buddies send hormonal signals or text messages to their sisters, your mitochondria, including the ones in your brain. These text messages travel "wirelessly" via your bloodstream and lymph system. But your gut has another old-school way of sending messages to the "second brain" in your head. The gut and brain communicate via the vagus nerve, which is the longest nerve of the autonomic nervous system and is equivalent to the landline or cable system in your home. The vagus nerve controls most of your autonomic (unconscious) bodily functions, such as heart rate, respiratory rate, digestion, and so on. The vagus nerve runs between the gut and the brain, snaking around the various organs in your body along the way.

When one part of your body needs to communicate with another, it makes a call along this landline to send the message. For many years, we all believed that the vagus nerve existed for the brain (the one in your head) to communicate with and give orders to the rest of the body, including the gut. That was what I was taught in medical school and believed for much of my career. But we now know that it is actually the other way around: for every nerve fiber leading from the brain in your head to your heart, lungs, and your gut, there are *nine* nerve fibers leading to the brain from the latter. There is therefore nine times as much communication going from the gut to the brain than there is going in the opposite direction.

To put it simply: *your gut buddies are the ones making the calls.* Not only that, but there are actually more neurons lining your gut to receive and interpret these messages than there are in your entire spinal cord. The 99 percenters are truly in charge of the way you act, think, and even feel. You female readers will recognize that "gut sense" you are so wisely conscious of. This is truly your sixth sense, the interpretation and integration of what is happening in your inner world, or the other you. Just to refresh your memory (this is the memory chapter, after all), both your gut bugs and the mitochondria that power your neurons were inherited from your mother. The sisters are always talking . . .

Gut sense works like any of your other senses. Your eyes, for example, are stimulated by photons of light that are converted into electrical signals that travel along the optic nerve to the back of your brain, where those signals are reassembled into the pixels that you "see." Similarly, sound waves vibrate tiny hairs in your ears, and, well, you get the idea. But unlike your other senses, which use cable or hard wires, your sixth sense uses not only the vagus nerve but also hormonal "cellular" systems to communicate.

With your gut buddies making so many calls to your brain, anytime the gut lining is breached and/or invaders sneak through this barrier and into your body, multiple chemical text messages are sent very quickly throughout your bloodstream and lymph system. As you learned earlier, the text messages are called cytokines, and the ones that alert the immune system and high command that an imminent threat or actual attack exists are called inflammatory cytokines. Over the last ten years, sophisticated blood tests have allowed me to measure these cytokines in my patients' blood every three months. As I reported at the American Heart Association,[3] these measurements first opened the window on how often lectins and LPSs routinely breach the gut wall defenses and both prompt local inflammation in your gut and invite the same inflammation into structures such as your heart, your joints, and your brain.

Inflammation in the brain (which is called neuroinflammation) is terribly damaging. This is reflected in the latest research, which shows that gut microbe–driven neuroinflammation results in the collateral damage of neurons that your brain immune system is sworn to protect and causes the cognitive decline that we think of as a normal part of aging. Neuroinflammation is now recognized as the true cause of the serious degenerative diseases such as Parkinson's, Alzheimer's, and dementia.

The connection between neuroinflammation and cognitive decline—and even degenerative disease—is so strong that my colleague Patrick McGeer has begun experimenting with administering low-dose anti-inflammatory medications (NSAIDs) to people who are at risk for developing Alzheimer's disease. Though his studies are promising, I worry about the long-term effects of NSAIDs on the gut barrier and prefer a protocol that will reduce inflammation at its source—the gut.

KEEP TROOPS AT THE OUTPOST

Research on mice clearly shows that certain changes to the microbiome lead to neuroinflammation and therefore cognitive decline. In one study, when aged mice were given an abundance of one type of bad bugs (from the Porphyromonadaceae family), they developed gut permeability and inflammation, and that led those poor old mice to have problems with spatial memory and an increase in anxiety-like behaviors.[4]

Even in humans, we can see how specific bacterial populations contribute to Alzheimer's disease. When my colleagues looked at the gut biomes of patients who were cognitively impaired and compared them to the bacteria in patients without such impairments, they found that the cognitively impaired group had an abundance of proinflammatory bad bugs (*Escherichia* and *Shigella*) and a reduction in anti-inflammatory gut buddies (*E. rectale*).[5]

Once again, over time, bad gut microbes, combined with the lectins you eat and their leakage through the gut wall with LPSs, break down the gut wall and set off an immune response that creates systemic inflammation.[6] These invaders are directly implicated in neuroinflammation and neurological disease. The neurons in the brain produce essential neurotransmitters such as dopamine, which stimulates neurons, and gamma-aminobutyric acid (GABA), which calms neurons down. Dopamine has the lofty job of regulating emotions, moods, and muscle movements, as well as sensations of pleasure and pain. Dysfunction of this critical neurotransmitter lays the groundwork for a number of diseases, most notably Parkinson's disease.

Neurons have long, spindly structures—axons and dendrites—that extend out from the cell body. The axons send messages to other neurons, and the dendrites receive messages from other neurons. Your neurons use these structures to create neural networks, and communication among these networks controls most of your thoughts, actions, and even movements.

What does this have to do with inflammation? Bear with me for one more moment while I tell you about another special type of cell called glial or microglial cells, which are like bodyguards or secret service agents for your neurons. Neurons are so important that they have their own handlers. You can think of neurons as the celebrities of your inner condo. And their bodyguards mean business. When they detect (via the inflammatory cytokine text messages sent from the border patrol in your gut wall) that invaders have breached the outer walls or the blood-brain barrier, the glial cells actually prune away the neurons' dendritic structures in their quest to protect against invaders. When this happens again and again, all that is left is the body of the cell, which now has no way of communicating with other neurons. If your neurons can't communicate with one another, of course you begin to suffer from problems with memory and cognition!

Some of my colleagues believe that dementia is caused by dysfunctional glial cells and are developing pharmaceutical interventions to stop glial cells from munching dendrites. But I (and many others) believe that these bodyguards are doing their job of guarding the neurons against presumed invaders. Perhaps they are doing their job *too* well. Think of it this way: a neuron is a central fort with outposts (dendritic structures) all over the periphery of the kingdom. A glial cell (which receives messages) finds out that the enemy (in this case, LPSs and lectins) is advancing, so it calls back to the central fort. The commander at the central fort says, "Get out of there! Come back to the fort!" The glial cells then commence efforts to protect the neuron (aka the fort) by cutting off those dendritic structures (outposts) until all that's left is the central fort and they don't have to worry about protecting the outposts.

So the underlying problem is not glial cells gone wild but rather the glial cells being triggered in the first place. Even worse, once the glial cells have gotten the neuron down to the cell body with no dendritic structures, they crowd around that cell body like the bodyguards they are. Sadly, they protect the nerve so well that even simple nourishment can't get in, and the neuron ultimately dies.[7] In Parkinson's disease, a typical finding of this disease is called a Lewy body, which is a dead neuron surrounded by glial cells. Glial cells mean well, but they are overeager and end up hurting the very thing they are trying so hard to protect.

When lectins and LPSs cross the gut barrier, not only do they travel through the bloodstream, but they also climb the vagus nerve from the gut to the brain.[8] There they are deposited into the substantia nigra,[9] an important structure in the midbrain that plays a critical role in motor planning. The neurons in the substantia nigra are dopamine-producing cells, and when inflammation causes a loss of neurons in this part of the brain, the result is Parkinson's disease. Indeed, as I explained in *The Plant Paradox*, humans who had their vagus nerve cut as part of now-antiquated ulcer treatments have only half the incidence of developing Parkinson's as folks with the nerve intact.[10] Think about that: if the phone lines are cut, lectins can't get in, but, more important, the brain remains oblivious to the fact that there's trouble heading its way.

Moreover, invading lectins and LPSs encounter large

numbers of neurons and their protective glial cells at the gut border. Now, here's where it gets really interesting. Parkinson's patients suffer from constipation, and my colleagues in neurogastroenterology (yes, there is such a discipline) suspected that something was amiss with the neurons in the gut wall that affect gut motility (bowel movements). Animal experiments found that when the gut wall was breached, Lewy bodies, those dead neurons surrounded by their bodyguards, could be found in the gut! Subsequent transcolonic biopsies in Parkinson's patients have confirmed the animal studies.[11]

So even this, the second most common neurological disease with more than 200,000 new cases diagnosed in the United States each year, stems from changes in the gut, not the brain. Want more bad news? Glutamate, an amino acid produced by your gut bacteria from glutamine, kills off dopamine-producing neurons.[12] Where do you get glutamine? From MSG, the flavor enhancer present in so many prepared foods. Don't see it on the label? The FDA considers it to be "GRAS" (generally regarded as safe), so its presence in a product is not required to be listed on a food label. If you see "natural flavors" on an ingredients label, run for your life (and your brain's life)! Would you believe that aspartame, which you may know as NutraSweet (the pink sweetener packet), converts to glutamate in your gut? Think twice next time you sweeten your coffee. By the way, aspartame was the preferred sweetener in diet drinks until very recently.

Pretty scary stuff.

But there is plenty of hope. Researchers at the UCLA Longevity Center have recently completed the first long-term, double-blind study of a bioavailable form of curcumin (found in turmeric), an anti-inflammatory polyphenol. After splitting a group of forty people into two groups, one that took curcumin and another that took a placebo daily, and correcting for age, sex, and other biological factors, the researchers found that the group taking curcumin had significant improvements in verbal memory, visual memory, and attention span after eighteen months.[13] Curcumin is one of the few known compounds that can pass through the blood-brain barrier to quiet neuroinflammation.

This is just the tip of the iceberg. Later in the book we'll talk about many more foods and supplements you can consume that will nourish your gut buddies and stave off cognitive decline and illness. But there are a few other issues to address first.

Multiple Sclerosis in the Gut

Multiple sclerosis (MS) is a debilitating disease in which the immune system eats away at the protective lining of nerves, called myelin. The resulting nerve damage disrupts communication between the brain and the rest of the body and/or within the brain itself. With everything you now understand about the immune system and this line of communication, would it surprise you to learn that even this disease has its origins in the gut?

We now know that there are specific gut buddies that produce molecules you've heard about before, called polysaccharides, that regulate myelination and demyelination (the creation and destruction of myelin). Without enough of these gut buddies producing enough polysaccharides, the immune system attacks the myelin, resulting in MS.[14] In one study, mice were given supplements of polysaccharide-producing bacteria. The result? Their myelin was better protected, making them less vulnerable to MS.[15]

Does this mean that nourishing that strain of gut buddies can help prevent or treat the disease in humans? Possibly. This is extremely promising for the millions of people suffering from MS, which is normally treated with drugs that suppress the immune system. As my good friend Dr. Terry Wahls has demonstrated, MS is clearly reversible by diet. Dr. Wahls reversed her own MS, which had left her wheelchair bound, by eating nine cups of vegetables a day and removing the vast majority of lectin-containing foods from her diet.

Recently I reported at the American Heart Association EPI/Lifestyle Scientific Sessions on 102 patients with biomarker-proven autoimmune disease, including several patients with MS, who went on the Plant Paradox program for six months. Ninety-five of the patients are now biomarker negative, symptom free, and off all immunosuppressive medications.[16] I firmly believe that in the near future we will be treating this and all other degenerative diseases exactly where Hippocrates suggested: in the gut.

WASH YOUR BRAIN EVERY NIGHT

Just when you thought you'd read the full horror story about gut invaders and their role in inducing neuroinflammation, there's more. For more than a decade, some of my colleagues studying Alzheimer's disease have been focusing on amyloids, which are dysfunctional proteins that stick together and form fibrous plaques around cells, after they observed that patients with Alzheimer's disease had accumulations of beta-amyloid in their brains.[17] The theory was that the brain produced the amyloids, which killed off neurons and caused Alzheimer's disease. Pharmaceutical companies have spent tens of billions of dollars on developing antiamyloid drugs that would allegedly treat and prevent Alzheimer's. Sounds pretty good, but all of those drugs have been a dismal failure. Why? Because amyloids aren't coming from the brain at all; they're coming from the gut.

Just as your gut bacteria create LPSs each time they divide and/or die, they produce amyloid when their own proteins die or become dysfunctional. They are called *shed molecules* because they literally split off from healthy bacteria and then travel through the body either by penetrating a vulnerable gut wall or by climbing the vagus nerve to the brain.[18] Once they reach the brain, they stimulate the brain to produce more amyloid, therefore becoming self-sustaining. In addition to disrupting normal cellular function, amyloid proteins then interfere with communication between cells.

Two things here. First of all, if you have a healthy population of gut buddies, they won't make amyloid proteins to begin with. And if they do, as long as the amyloids don't make it out of the gut, they'll never cross the blood-brain barrier. Our bodies are accustomed to dealing with both LPSs and amyloids. These "shed molecules" do not automatically cause problems just because they exist, and they do not automatically need to cause problems as we age, either. But problems do occur over time as the gut barrier becomes increasingly permeable.

Second, even if they do reach the brain, amyloids don't automatically cause damage; scans of many patients with no evidence of Alzheimer's show the presence of amyloid proteins. The all-important distinction is whether the plaques get washed out of the brain or are allowed to remain there, where they trigger the production of more and more amyloid, becoming self-perpetuating and eventually accumulating in the brain and contributing to Alzheimer's and other forms of neurodegeneration.

THE BRAIN'S HOUSEKEEPING SYSTEM

In every part of your body, the spaces in between cells must be regularly swept clean to get rid of junk and debris. Throughout most of your body, your lymphatic system is recognized as doing this important cleanup work. Lymph, a clear fluid containing proteins and white blood cells (your cops), flows through the body and drains away any garbage that hasn't been properly recycled. But until recently, no one knew whether or not a similar process occurred in the brain. The blood-brain barrier was thought to keep lymph from reaching the brain.

A few years ago, researchers discovered a system that allows cerebrospinal fluid (that clear fluid that we tap when we stick a needle into your spine) to flow through the brain, cleaning out the spaces in between cells, just as lymphatic fluid does in the rest of your body. This is called the *glymphatic system*. To make room for the fluid to wash out your brain, your cells actually shrink in size when you are in a deep sleep. This allows the full "brain wash" process to go twenty times as fast when you are in deep sleep as when you are awake and helps to explain why a good night's sleep is so restorative.[19] When you get an adequate amount of deep sleep, you literally wake up with a refreshed and rejuvenated mind that has been swept clean of junk and debris.

So as long as you're going to bed at a reasonable hour, you should be getting the full brain-wash treatment every night, right? Unfortunately, this is not the case. You read earlier that when I was a boy I wasn't allowed to swim for an hour after I ate. That was because of my mother's fear that all of my blood would rush to my stomach to digest whatever food I had eaten, leaving no blood left to flow to my muscles and give me the strength to swim. Believe it or not, this same principle relates to your sleep.

The glymphatic system is most active during the specific stage of deep sleep that happens very early in the sleep cycle. And the glymphatic system, just like your digestive system, requires a great deal of blood flow. This means that if you eat too soon before going to bed, your blood will all flow to your gut to aid in digestion and will not be able to reach your brain to complete the all-important brain wash. In this case, it's your brain—not your muscles—that suffers from lack of blood flow while your body is focused on digesting your food. And the results for you can be almost as serious as drowning. If your glymphatic system cannot wash out your brain effectively, you end up with a buildup of amyloid and other toxins, including lectins and LPSs, in the brain, and you already know what that causes. Sadly, this "wash cycle" is probably the single most overlooked and misunderstood aspect of neurodegenerative diseases.

Luckily, there is a simple solution: leaving as big a gap as possible between your last meal of the day and your bedtime. As my friend and colleague Dr. Dale Bredesen, the author of *The End of Alzheimer's: The First Program to Prevent and Reverse Cognitive Decline*, has shown, the minimum amount of time between finishing your last meal and going to sleep should be four hours. Remember, the one-hour rule was a bit of an old wives' tale that held some truth. Unfortunately, it takes more than just one hour for your body to fully digest your dinner so that you will have adequate blood flow to wash out your brain. This means that if you go to bed at eleven, you should eat nothing after seven o'clock. There is a lot of wisdom in the old maxim "Eat like a king at breakfast, a queen at lunch, and a pauper at dinner." Unfortunately, our culture promotes the opposite—eating a large, late dinner (or, even worse, a snack right before bed) that forces your body to spend its efforts digesting your food during the night, when it should be focused on cleaning out the gunk in your brain.

I understand that it is not always possible to leave a full four hours between dinner and bedtime. Many of my patients work long hours or go out to late dinners with friends, colleagues, and/or clients, and then they don't want to have to stay up for four hours while they finish digesting. To get the full cleansing effects without sacrificing your schedule or sanity, I have incorporated a "brain wash" day into the Longevity Paradox program. Once a week (or more), when it is convenient for you, you will skip dinner to make sure your blood can flow freely to your brain as soon as you fall asleep. Think of it as a weekly deep clean when you're too busy to tidy up on a daily basis. You'll also get all of the benefits you read about earlier from temporarily stressing your cells with this brief intermittent fast.

Obesity and the Brain

Most likely because of its impact on inflammation throughout the body, obesity is a major risk factor for dementia. A new study out of University College London that looked at more than a million people in the United States and Europe over a period of thirty-eight years found that people with a high body mass index (BMI) were significantly more likely to develop dementia than people who had a lower BMI.[20] In fact, those who had central obesity (in the gut) were the most affected. So remember my old saying: Fat in your gut, you're out of luck! This speaks further to the dangers of living in a 365-day growth cycle.

You'll read much more about what actually causes weight gain in the next chapter. (Spoiler alert: it's your gut bacteria.) But for now rest assured that the Longevity Paradox program will help you safely lose weight to improve the health of your body and mind.

EAT FOR YOUR BRAIN

Amazingly, there is one food that can counteract nearly all of the damage we've just discussed, dramatically reducing your likelihood of developing dementia or other neurological diseases, and it's one that most people enjoy: good old-fashioned extra-virgin olive oil.[21] I've often said that the only purpose of eating any food is to get more olive oil into your mouth. A staple of the Mediterranean diet as well as in the diets of three of the longest-lived communities on Earth, olive oil has long been heralded as a cure-all for health and longevity. But new studies on olive oil and brain health are so compelling that it should also be used as a therapy for avoiding or slowing the development of dementia.[22]

I'd be happy to write you a prescription for olive oil, but hopefully you won't need it. Just head to your local Costco or Trader Joe's and make sure you're buying a first-cold-pressing bottle. And no, it doesn't have to be from Italy. There are many good and reasonably priced American brands, such as California Olive Ranch and Bariani.

But why is olive oil so good for you, and how does it specifically impact the brain? Well, we've known for a long time that olive oil has anti-inflammatory properties, due to its high level of polyphenols.[23] Remember, your gut buddies convert polyphenols into anti-inflammatory compounds. Moreover, they suppress the ability of gut bugs to produce the vessel-damaging compounds I mentioned previously that are formed from animal proteins, called TMAOs.[24] That alone may explain why consumption of olive oil is associated with a reduced risk of Alzheimer's. But we now know that olive oil kick-starts autophagy, the highly beneficial cellular recycling process. Indeed, mice that are fed diets rich in olive oil have higher levels of autophagy than do mice that are fed a conventional diet. Furthermore, olive oil–consuming mice perform better on memory and learning tests than their non-olive oil–fed peers.[25]

But it's not just autophagy. Mice that are fed olive oil also have reduced levels of amyloid plaques in their brains. How can this be? It turns out that an olive oil–enriched diet stimulates neurons within the brain stem to release a hormone called glucagon-like peptide-1 (GLP-1), which decreases blood sugar levels and is associated with many benefits, including weight loss and a lower risk of developing hypoglycemia.[26] Because of GLP-1's effects on blood sugar levels, GLP-1 supplements are also a promising treatment option for type 2 diabetes.

But GLP-1 also protects synaptic activity (those important connections between dendrites and axons) in the brain from amyloid toxicity.[27] It does this by stimulating the production of brain-derived neurotrophic factor (BDNF),[28] an enormously beneficial protein that promotes the growth of dendrites and axons and supports their connectivity. This means that olive oil actually helps your neurons repair themselves after glial cells have bitten away their dendritic structures! But even if your neurons have not been damaged by inflammation, BDNF supports the growth of new neurons, leading to better long-term memory and cognition.

That's right: simply consuming olive oil reduces inflammation, stimulates autophagy, supports the growth of new neurons, helps regrow dendritic structures that form neural networks, and protects your brain from the negative effects of any amyloids that don't get washed out at night. As I said, why bother eating anything that you don't pour olive oil onto?

In a four-year study recently published from Spain, 447 adults aged 67 were assigned to one of three diet groups: one group was instructed to consume a liter of olive oil per week (that's about 9 to 10 tablespoons a day); another ate 30 grams of walnuts a day; and the third group consumed a similar number of calories but followed a low-fat diet. Based on brain function tests administered before the study began and after four years on the diets, the low-fat group experienced a significant reduction in memory and cognition, while the nut-eating group showed significant improvements in memory and the olive oil group experienced significantly improved overall cognitive function.[29] Attention, class: consuming nuts and olive oil makes you smarter! From what you've learned so far, the bottom line is clear: both of these foods change your gut buddies for the better.

The benefits of olive oil explain why the Mediterranean diet is consistently associated with healthier aging despite the prevalence of grains. The olive oil in the Mediterranean diet offsets the negative effects of the grains' lectins and can help preserve your brain well into old age. This has been proven in studies that look at how the Mediterranean diet impacts gut health. After one group of monkeys ate a Mediterranean-style diet and another group ate a Western diet for a period of two years, the ones that ate the Mediterranean diet had significantly higher microbiome diversity and a better ratio of gut buddies to bad bugs.[30] It's very simple: if you want a healthier microbiome (and therefore a healthier brain and body), eat plenty of olive oil!

In another study, researchers in Scotland followed a group of more than four hundred people between the ages of 73 and 76, analyzed their diets, and then scanned their brains regularly over a period of three years to see how their dietary habits impacted their brain health. They found that the people who ate more olive oil and less fried food and red meat had about *half* the rate of brain shrinkage considered normal for their age group. They concluded that the subjects' diet provided long-term protection to the brain.[31]

These results are pretty remarkable, but if you think about everything you've learned so far, it makes perfect sense that a diet high in olive oil not only would result in the death of fewer neurons than a diet high in inflammatory fried foods and animal protein but would actually make new brain cells grow. Another loss for this boy from Nebraska, but a win for aging brains everywhere.

SOMETHING'S FISHY

The Mediterranean diet has another component that always drew my attention as I researched the omega-3 index and especially after my recent visit to Acciaroli, the town with the largest percentage of centenarians in the world. That component is consumption of small fish. A recently published study that followed nearly 400,000 men and women for sixteen years showed that those with the highest fish and long-chain omega-3 fatty acid intakes had significantly lower total mortality and lower cardiovascular and respiratory mortality. Even more striking, the women with the highest omega-3 intake had nearly 40 percent less Alzheimer's disease mortality! Spoiler alert: fried fish consumption didn't produce the same results.[32] So back off on those fish-and-chips!

Even more striking are the recent findings from Dr. Daniel Amen using brain SPECT (nuclear imaging) scans: patients with higher omega-3 indexes had increased perfusion (blood flow) in regions of the brain associated with learning, memorizing, and avoiding depression. Those with the lowest levels of omega-3s had the most diminished perfusion in the same areas.[33] After following more than a thousand women as part of the Women's Health Initiative Study over eight years and looking at the same omega-3 index, a study published in the journal *Neurology* reported that women with the highest omega-3 intakes had the biggest brains and the biggest memory areas (the hippocampus) compared to those with the lowest levels.[34] Your mother was right: fish is brain food!

So what are the best sources of omega-3s? In my patient population, the highest omega-3 indexes without supplements are found in those who are daily sardine or herring eaters. Remarkably, my patients from Seattle and Vancouver who eat wild salmon almost daily have never achieved high levels without additional supplementation. All of which takes us back to those Acciaroli centenarians and our old friends the Kitavans: they eat primarily anchovies and other small fish. Hate sardines and herring? Not to worry, there are plenty of other ways to get your long-chain omega-3s, and no, vegans, it's not from flaxseed oil.

Eat Your Greens!

You already know that your gut buddies love it when you eat leafy greens. And when you do, they return the favor by powering your brain. A major study at Tufts University on almost a thousand participants aged 58 to 99 over nearly five years showed that eating just one serving of leafy green vegetables per day slowed the aging process in the brain. Specifically, after adjusting for age, sex, education, participation in cognitive activities, physical activity, smoking, and seafood and alcohol consumption, the participants who regularly ate leafy greens slowed their rate of cognitive decline by eleven years.[35] Imagine making your brain eleven years younger just by eating greens every day! That's exactly what you can accomplish on the Longevity Paradox program.

YOUR GUT BUDDIES ARE YOGIS

It comes as no surprise that meditation and yoga have many positive effects on the brain. But I believe that none of these changes actually occurs directly in your brain. Rather, your gut buddies really like it when you practice meditation and yoga, so when you do, they return the favor by improving your brain for you.[36] Some of my patients don't like the idea that tiny one-celled organisms can control the mood, emotion, and willpower of a "higher" creature such as a human being. They want to believe that they are in charge of their own brains. Well, I'm sorry, but you're going to have to get over it. You are nothing more than a transport vehicle and condominium for your microbiome. You might as well give them what they want.

The connection between yoga and/or meditation (yoga induces a state of meditation) and the gut all comes down to stress. Unhealthy stress levels (not the hormetic levels that are beneficial) affect the gut, promoting a higher population of bad bugs than gut buddies and altering gut permeability.[37] In fact, extreme stress in humans leads to exactly that: bad bugs and a leaky gut.[38,39]

In an experiment on mice, when the mice were stressed, their healthy gut buddies were reduced and their levels of bad bugs increased, resulting in more inflammatory cytokines produced in the gut.[40] And in a human study that examined the composition of fecal microbiota in forty-six patients with depression and thirty healthy controls, there were significant differences between the two groups, with increased populations of proinflammatory bad bugs and decreased population of gut buddies in the patients with depression.[41] Your gut buddies are in charge of your health, and your mental health is no exception.

Think of it this way: if there is a lot of tension in your inner condo or a lack of the foods your gut buddies need, they won't enjoy living there and will move on to greener pastures, allowing bad bugs to take over. Then things quickly go downhill. The bad bugs don't produce as much of the beneficial short-chain fatty acids and hormones that your gut buddies made for you. For example, your gut buddies produce norepinephrine, which works in the brain by increasing alertness, focus, and attention, and serotonin, the "feel good" neurotransmitter that increases feelings of well-being. Other gut buddies produce GABA, which, as you read earlier, calms neurons, promoting feelings of ease.

So it goes both ways. Stress alters the gut biome, and changes to the gut biome create feelings of stress. No wonder so many of my patients are on antidepressant and antianxiety medications when they first arrive in my office. Their gut buddies have sadly (pun intended) left the building.

Of course, the number one benefit of meditation/yoga is a reduction in stress. This in and of itself creates positive changes to your microbiome, which is what really provides all the cognitive benefits of these practices. Some of these benefits can help prevent Alzheimer's, while others have the potential to greatly extend your life. Studies on mice show that microbes in the gut are essential to increasing the ability to deal with a stressful situation appropriately.[42] In fact, transplanting nonanxious rodents with the microbiota of anxious rodents produced—you guessed it—more anxious rodents![43] So are my patients' gut bacteria actually creating the feelings of anxiety that they suffer from? Many of my colleagues and I believe so. In other words, your gut microbiome composition has more to do with your anxiety levels than with how you "feel" in your brain. Although, of course, external circumstances may play a role in creating your feelings, my colleagues at University College, Cork, in Ireland concluded, "Resilience to environmental stress seems to be heavily influenced by microbiotal composition."[44]

At the University of California, Davis, researchers compared thirty participants on a meditation retreat at the Shambhala Mountain Center in Colorado with people on a waiting list for the retreat and corrected for age and general health. The participants who attended the retreat meditated for six hours per day for three months. At the end of three months, the meditators had on average a full 30 percent more activity of the enzyme telomerase than the control subjects did.[45] Telomerase is the enzyme that activates telomeres, as discussed earlier, the length of which may be a marker for longevity. And what causes an increase in telomerase activity? Your stem cells, which are activated by your gut buddies. All that meditation made the participants' gut buddies happy, and they told the stem cells to increase telomerase so they could stay in their pleasant, stress-free abodes for the long term. Sounds hokey, but it's true.

Another study looked at healthy male volunteers from the Indian Navy who were divided into two groups—one that practiced yoga and meditation for six months and one that practiced routine physical training exercise instead. In the yoga and meditation group, glutathione activity increased significantly after six months.[46] Glutathione is the principal intracellular antioxidant that protects you from aging. And guess what powers its production? Butyrate, the beneficial amino acid produced by your gut buddies![47] In this case, the gut buddies enjoyed the manipulation of the gut by yoga, so they produced more butyrate, which increased glutathione levels to protect against aging.

Additional studies show that yoga and meditation prevent age-related degeneration by increasing BDNF levels and therefore protecting your neurons from the effects of inflammation.[48] And where does BDNF come from? Your gut buddies make it.[49] Other studies show that yoga practitioners have significantly greater gray matter concentration in both the prefrontal cortex and the hippocampus and experience fewer cognitive failures.[50] This BDNF boost explains why even though most people lose brain mass as they age, long-term meditators do not.[51] As a result, they have improved verbal and visual skills and enhanced global awareness and attention.[52] A brand-new study at Massachusetts General Hospital shows that other forms of exercise also benefit the brain by boosting BDNF production and cleansing the brain, creating a better environment in which new neurons can grow.[53] More on this later.

Face it: you are a home for your bacteria, and they like it when you practice meditation and yoga and want you to stick around into ripe old age to keep doing it. So, try it! See how they, and hence you, like it. But a word of warning: meditation cannot cure everything. If you don't eat for the 99 percenters and protect your gut lining, meditation can do only so much.

Let me share a recent example. A Japanese meditation expert named Aiko Watanabe, who is Deepak Chopra's representative and translator in Japan, recently came to see me. She suffered from lifelong debilitating rheumatoid arthritis, which she had tried to treat with diet and meditation. But her best efforts simply and tragically hadn't worked. By the time she came to me, she'd already had two knee replacements and was on two immune suppressants and in such terrible pain she was essentially bedridden. She had been following a macrobiotic and Ayurvedic diet that included a lot of brown rice, a major source of lectins that were broaching her gut lining and causing inflammation. Since starting my protocol, she's been off of her medications and pain free for months. She recently walked into my office to show me the remarkable changes. Remarkable? Yes. But you, too, can achieve remarkable results when you restore the proper functioning of your gut.

This is not to say that meditation and yoga won't help combat inflammation. But I've found that unless you couple these practices with a lectin-limited diet that prevents disruption of the gut wall and supplement with foods that feed your gut buddies, you are unlikely to experience the full benefits. Make all of these changes, and I promise they will thank you with a more youthful body both inside and out—which leads us to our next chapter . . .

Chapter 7
Look Younger As You Age

Though my patients' primary concerns about aging are generally related to developing disease, memory loss, fatigue, and limited mobility, let's face it: we all have a secondary set of concerns about aging—and that is what we see in the mirror!

Some people may think that being invested in your appearance as you age is vain, but I disagree—after all, the way a person looks affects his or her self-esteem and therefore his or her mental well-being, which is a huge predictor of health outcomes. And with 80 percent of the population gaining an average of twenty-three pounds or more since the 1980s and putting themselves at a higher risk of all the diseases we have mentioned so far, weight is much more than simply a cosmetic issue.

But another reason to focus on the external is that what you see on the outside of your body is a direct reflection of the state of affairs inside. So if you've been gaining weight over the years and notice your skin getting thinner, more deeply wrinkled, or discolored as time goes on, you can be sure that something is rotten in your core. And although some superficial changes are inevitable, it *is* entirely possible to maintain soft, supple skin, reverse the damage that's already been done, lose weight, and get into even better shape as you age.

As you learned earlier, the lining of your gut is equivalent to the surface area of a tennis court in size, covering vastly more square footage than your external skin. Yet in many ways it functions just like your external skin: it protects you from "the elements," it feels and senses (remember your sixth sense?), and it separates what you swallow and your microbiome from the rest of you. Have you ever wondered why the skin of most older folks seems paper thin? That thinning is directly connected to the gut wall weakening and allowing more and more unwanted LPSs and bad microbes access into the blood. As your inner skin begins to break down and thin over time, that aging and thinning are reflected in the skin you see in the mirror every day.[1]

Think about it this way: when you're looking to buy a home, the first thing you see when you pull up at the address on the real estate listing is the exterior of the house. You can tell a lot about a house from the exterior. Is it well cared for, with neat landscaping, fresh paint, and no visible signs of water or other damage? Then it's pretty safe to assume that the inside of the house will also be clean and welcoming. But if there are overgrown bushes and weeds outside with visible mold, broken windows, peeling paint, and rotting wood, the inside of the house is probably just as run down. Any real estate agent will tell you this is true: to some extent, you *can* judge a book by its cover.

I am evidence of this myself, and so was Edith Morrey ("Michelle") when I met her at age 90 and mistook her for 65! People are often surprised to learn that I am 68 years old—and that I haven't made any dermatologists (or plastic surgeons) rich along the way. In fact, I've never gone to one! I simply fixed my diet, healed my gut, and—voilà—my skin got smoother and my weight dropped dramatically. The same thing can happen for you.

Remember, 99 percent of your genes are not *your* genes—they belong to your gut buddies. And your gut buddies are in charge of your beauty from the inside out.

YOUR GUT BUDDIES AND YOUR WEIGHT

Recall the maxim you read earlier: *You are not what you eat; you are what your gut buddies digest.* When it comes to your weight, this is a crucial truth that is far too often overlooked. A revealing new study from the University of Chicago Medical Center[2] confirms the essential role of gut microbes in weight management. When researchers analyzed the gut bugs that live in the upper gastrointestinal tract rather than those that reside in the large intestine (where much gut research is focused), they found an important clue as to why we gain weight as we age: a strain of bacteria lives in the upper GI tract that helps you rapidly digest and absorb high-fat foods. The scary thing is that they multiply to keep up with the job—so the more fat that is present, the more of these bugs you will have. If you eat a standard Western diet, over time you will gain more and more of these bacteria. The result is that they quickly and efficiently digest all that fat and send it right into your body. Once again, it's not the calories you eat that makes the difference; it's the calories "they" make available to you.

Say it with me: *You are not what you eat; you are what your gut buddies digest.* In a study by the same team to prove the mechanism, germ-free mice that did not possess these fat-digesting gut bugs were fed a high-fat diet and did not gain any weight. Where did all that fat go, then? They pooped it out, quite literally. They had elevated lipid (fat) levels in their stool. Even more compelling, when those germ-free mice were later exposed to the fat-eating bacteria and fed a high-fat diet yet again, they did gain weight because this time their gang members digested and passed that fat on to them.

What's that saying? *You are not what you eat; you are what your gut buddies digest.* No, the other one. *When you take care of your gut buddies, they will take care of you.* That's it! Eating a high-fat diet feeds the bad bugs that want to take over and let their home rot and starves the gut buddies that want to keep you young and lean. You need to keep your microbiome fueled with the foods they love to maximize your curb appeal, and these are the exact foods you'll be eating on the Longevity Paradox plan.

ENDOCRINE DISRUPTORS ARE MAKING YOU FAT

You read earlier about the dangers of hormone disruptors. In addition to promoting early growth and puberty, these disruptors cause adults to continue to gain weight as they age. This is one of the leading causes of the obesity epidemic but one that few physicians are identifying in their patients who struggle with their weight.

The main hormone these chemicals disrupt is estrogen, which we normally associate with women but is actually present (in varying amounts) in both sexes. In women of childbearing age, estrogen's main job is to tell the cells to store fat to prepare for an upcoming pregnancy. This was crucial back when we lived in a yearly cycle of growth and regression. During the growth cycle, a woman would gain weight so that she (and her baby) could live off that stored fat during the leaner times to come. In fact, that's why very thin female athletes sometimes don't menstruate. Their internal fat storage counter doesn't think there are enough fat deposits to nourish a baby and won't risk wasting a precious egg.

But now we live in a 365-day growth cycle, and there is quite frankly no need for women to store fat ahead of time whether they are planning to become pregnant or not. Food is always available for both mother and baby. But when toxins in our environment mimic estrogen in the body of a man or a woman, our cells get the message to store fat regardless of whether or not we are even biologically capable of becoming pregnant. This is why some girls are starting puberty at the age of eight and many of my male patients first come to me with "man boobs" and a huge gut that looks like a pregnancy belly.

The idea that minute amounts of these estrogen-mimicking compounds could cause trouble has been ridiculed even by those at the EPA and the FDA, whose job it is (or, sadly, was) to protect the unsuspecting consumer from such things. But the cumulative effect of the minute amounts of estrogen-like compounds we regularly absorb from our environment is more powerful than the hormone itself would be.[3] Instead of hooking up to an estrogen receptor on a fat cell, delivering their message, and then leaving the way the regular estrogen hormone does, estrogen-like compounds attach to the receptor and never leave, keeping the fat cell permanently switched on to keep storing fat. This further disrupts normal cell messaging. Thanks to the effects of these compounds, men, women, boys, and girls are constantly storing fat for a nonexistent upcoming pregnancy!

To avoid unnecessary weight gain as you age and the health problems that come along with it, it is essential to avoid these hormone disruptors. Here are some of the most common and harmful ones.

Bisphenol A (BPA)

BPA has been used commercially since 1957 in the synthesis of plastics. It is also used to line pipes and to coat the insides of many food and beverage cans. In the past decade or so, scientists caught on to the estrogen-mimicking effects of this chemical. The Food and Drug Administration has ended its authorization of using BPA in baby bottles and cans that store baby formula. BPA is banned completely in Canada and Europe, but in 2015 a lawsuit in the United States attempting to force the FDA to ban BPA was defeated, thanks to a large donation by the American Chemistry Council to congressional campaigns that opposed the bill.[4] Two years later in 2017, the European Chemicals Agency announced that BPA should be listed as "a substance of very high concern" due to its properties as an endocrine disruptor.

As a result of public awareness (and therefore market abandonment) of BPA, many companies have chosen to stop using it, but others have not. And its "safe" replacement, bisphenol S (BPS), was recently found to have the same endocrine-disrupting properties.[5] I recommend limiting your exposure as much as possible by:

- Avoiding canned foods (buy frozen foods if you can't get them fresh), unless you see a label that says "BPA-free lining"
- Using glassware instead of plasticware for food storage

- Never heating food in plastic in the microwave (heat causes BPA to leach out of plastics into food)
- Using glass or stainless-steel water bottles instead of plastic
- Making sure that any plastic toys (especially for young children, who will put them into their mouths) specifically state that they are BPA free
- Asking for cash register receipts (which contain BPA) to be placed into your bag or just thrown away

Phthalates

These synthetic compounds that are used to soften plastics began to appear in the early twentieth century and are now ubiquitous in wall coverings, vinyl flooring, the gloves you wear to wash dishes, the trays used to package meat and fish, the plastic wrap you cover leftovers with, and even the toys your children play with. Yikes. And thanks to their presence in plastic wraps and plastic containers, phthalates are omnipresent in our foods. Phthalates also act as solvents in perfumed items, so they turn up in hair sprays, lubricants, insect repellents, and thousands of other household and personal care products.

Animal and human studies have consistently connected phthalates to endocrine disruption. In one study, they were associated with smaller-than-usual testicles in rats.[6] In humans, the presence of highly concentrated amounts of phthalate metabolites in men's urine has been associated with damage to the DNA in sperm.[7] Being exposed to these chemicals at a young age may be associated with premature breast development in girls.[8] And babies whose umbilical cords reveal higher-than-normal exposure to phthalates are more likely to be born prematurely.[9]

These compounds lock onto estrogen receptors in the brain and permanently attach to the thyroid hormone receptors on cells, blocking the real thyroid hormone from delivering its message. I see this all the time in my patients who are producing plenty of thyroid hormone yet have symptoms (including weight gain) of hypothyroidism. Although researchers in European countries, Canada, and China have conducted surveys in an effort to establish how much of this class of chemicals is in their food supply, the first US study[10] did not take place until 2013. It looked at a relatively pristine upstate New York population and found that the major sources of phthalates in humans were obtained from grains, beef, pork, chicken, and milk products.

Gee, I wonder why so many people who come to me who are on a steady diet of whole grain foods and boneless, skinless chicken breasts (wrapped in phthalate-laden food packaging) are tired and fat with thinning hair but have been assured by their doctor that their thyroid hormone levels are normal. Sure, they are making plenty of thyroid hormone, but it can't communicate with their cells because phthalates are in the way. Lo and behold, when we remove the phthalate-laden foods from their diets, these patients lose weight and gain energy and start feeling (and looking!) decades younger.

To avoid phthalates:

- Avoid most grains, conventionally grown meats, and dairy products.
- Use glass or stainless-steel water bottles instead of plastic.
- Never heat food in plastic in the microwave.
- Read labels of all personal care products. Make sure they specifically say "phthalate free."

Supplement with Sunscreen

Many conventional sunscreens are loaded with phthalates and other hormone disruptors. When buying sunscreen, look for products with a titanium oxide or zinc oxide base without preservatives such as parabens. But it's even better to eat your sunscreen by consuming plenty of vitamin C.

There is very good evidence that if we have plenty of vitamin C in our bodies, it will prevent solar damage to the skin. I take a time-released vitamin C supplement of 1,000 milligrams twice a day and have for years, and I rarely see the effects of the southern California sun on my skin. Humans are actually the only animals besides New World monkeys and guinea pigs that don't make their own vitamin C and need to rely on foods (or supplements) to get enough of it. Yet we do have all the necessary enzymes except the final one to produce vitamin C. So what gives?

Evolutionary biologists believe that when we evolved, we had so much vitamin C in our diets that the process of manufacturing it was "edited out" of our genes. Though eliminating our ability to produce vitamin C saved us from wasting energy and using that extra energy to store fat then, now it gets us into trouble when we don't consume enough of it. In 1966, the Nobel Prize winner Linus Pauling discovered our inability to make vitamin C after observing that the human body uses vitamin C to repair cracks in collagen. Collagen is the most abundant protein in our bodies, and is used to build connective tissue, skin, and blood vessels—think of it as like the rebar in concrete. When collagen breaks and isn't repaired well because of insufficient levels of vitamin C, we get wrinkles. The same is true of UV damage: the sun's rays harm your skin by breaking down collagen; vitamin C can knit it back together again—but only if you have enough of it.

So: vitamin C supplements are a beauty food. The problem with them is that they are water soluble, so you excrete vitamin C rapidly in your urine. There is also an upper limit to the amount of vitamin C you can absorb, and if you swallow more, not only will you not absorb it, you're also likely to get diarrhea as your body excretes whatever it can't absorb. Since animal studies suggest that you need plenty of vitamin C to maintain vital, healthy skin and blood vessels, a time-released vitamin C supplement taken twice a day should provide a lot of the sunscreen that you need.

Arsenic

This well-known poison is also an antibiotic that kills your gut buddies as well as a hormone disruptor. Talk about a triple threat! But somehow it continues to be used to color sickly-looking chicken a beautiful blushing pink.

Arsenic is a great example of a toxin that is *not* beneficial in smaller doses. It will kill you at a dose of 100 milligrams, and smaller doses will simply make your life shorter and less pleasant instead of making you stronger.

To avoid arsenic:

- Stop eating conventionally raised chicken. It is not a health food.
- Avoid most grains, particularly rice, as they contain arsenic.

Azodicarbonamide

Azodicarbonamide (try saying that three times fast) is another endocrine disruptor[11,12] that is used as a foaming agent in the manufacture of synthetic leather products, carpet underlay, and yoga mats and is also used to bleach flour and condition dough. Most fast-food restaurants, including Wendy's, McDonald's, Burger King, and Arby's, use it in some or all of their bread products although the use of azodicarbonamide in bread has been banned in the European Union[13] and Australia. In this country, Subway has voluntarily eliminated it from its products.[14]

Azodicarbonamide has been shown to provoke asthma and allergies,[15,16] as well as suppress immune function,[17] particularly when it is heated or baked (which of course normally happens to bread). Additionally, this chemical has been shown to break down gluten into two of its individual proteins, gliadin and glutenin, making them more immediately available and therefore more immediately irritating to your gut lining. This is one time when efficiency is not a virtue.

To avoid azodicarbonamide:

- Never eat fast food.
- Avoid most grains.
- If you do eat bread, choose organic fermented varieties (such as sourdough) and for now only when traveling outside of the United States, owing to the presence of the herbicide Roundup

in most of our wheat. But, spoiler alert! Roundup is now approved for use in Europe. Soon nowhere will be safe.

Blue Light

In addition to the chemicals that disrupt our hormones, we are constantly exposed to artificial blue light, which does the same thing. How? This goes back to our cyclical nature. For millennia, our ancestors sought and ate food in response to changes in daylight, specifically to the blue-wavelength spectrum of daylight. And our bodies are still wired to respond to these signals. Long days and short nights tell your body that it is summer, which stimulates your body to seek out food, especially the sugars in fruit, to store fat for the upcoming winter, when food will presumably be in shorter supply. Conversely, short days and long nights tell your body it is winter and you should seek less food because it is scarce and requires more calories to obtain and you can instead just live off the extra fat you've been storing all summer.

Hunting or foraging for food when little is available makes no sense because you would expend more calories than you are likely to obtain. Your body hates waste. So instead of seeking food in winter, we are designed to burn the fat we've acquired. The hormone leptin, which makes us feel full, turns on the signal that we should stop eating and start burning our stored fat during winter. This seasonal cycling between the use of glucose (food) for fuel and the use of stored fat for fuel is termed *metabolic flexibility.* And the instructions for this cycling are mediated by the blue spectrum of light. More blue light tells us to consume more glucose, while less blue light tells us to burn stored fat.

Sounds pretty simple, right? But our lives are now completely dominated by blue light. Televisions, cell phones, tablets, other electronic devices, and even most energy-saving lightbulbs emit light in the blue range of the spectrum, which is also known to interfere with sleep. Blue light suppresses the production of melatonin, the hormone that helps you fall asleep, and the resulting sleep deprivation is associated with obesity.[18]

Blue light also stimulates the hormones that make you feel hungry (ghrelin) and awake (cortisol), which of course adds to weight gain. And because our genetic programming associates blue light with daylight, this constant exposure tricks our bodies into thinking it's perpetual summer and that we should keep packing on the pounds in anticipation of the leaner days of winter, which of course will never arrive. Who knew that light could wreak such havoc on our bodies? For all of these reasons, I recommend that you minimize your exposure to blue light, especially in the evening, as much as possible.

To avoid blue light:

- Turn off all electronic devices when the sun goes down.
- Is that impractical? Put on a pair of blue-blocker glasses as soon as the sun goes down, and wear them until you go to bed.
- Use red night-lights so you can see where you're going if you get up during the night.
- Install an app such as f.lux on your computer to reduce the amount of blue light your screen emits; activate the nighttime lighting mode in the settings on your cell phone, tablet, and computer.

In addition to avoiding these harmful endocrine disruptors, in the Longevity Paradox program you'll return to the cyclical nature your gut buddies evolved in, which will make them happy and inspired to keep you trim, energetic, and young.

YOUR SKIN IS YOUR GUT LINING TURNED INSIDE OUT

We've already talked about the fact that your gut lining is, in essence, your skin turned outside in. What goes on in there comes back around to the surface of your skin. But just as your gut lining is in constant contact and communication with the gut microbiome, I would be remiss to leave out the trillions of bacteria living on your skin, especially in a book about aging (or about *not* aging, as the case may be). These bugs are commonly referred to as your *skin flora* and include more than a thousand different species of bacteria; together with the microbiome, they make up your holobiome.

Before you put down this book for your morning shower, give these guys (I'll call them your skin buddies) a chance. Like your gut buddies, they view your skin as their home, and as long as you take good care of them, they will take excellent care of you. In this case, taking care of you means keeping your skin soft and supple as you age. Your skin buddies have sworn to protect their home (your skin) at all costs. They will fight to the death against invaders such as bad bugs, mold, and fungi.[19]

Different skin buddies act to protect your skin in varying ways. One species of bacteria secretes antimicrobial substances that help fight pathogenic invaders, while another uses the skin's lipids to generate short-chain fatty acids to ward off microbial threats. (Yes, secretion of too much of these lipids by your skin buddies causes oily skin.) Another species secretes lipoteichoic acid (LTA), which prevents the release of inflammatory cytokines so that your skin does not become inflamed.[20] To maximize your skin's protection, it is therefore essential to have a richly diverse skin bacterial population. Your skin flora is meant to be even more diverse than your gut biome.

But over the years, we have done our skin buddies a horrible disservice. We assumed that having bacteria on our skin was a bad thing without paying any attention to what those bugs did for us, so we started to kill them off in droves with antibacterial cleansers and soaps. As a result, we have far less diverse populations of skin buddies in the United States than people in countries where they do not use these harmful products.[21] What's that? You think that we need antibacterial cleansers to keep away bad bugs and avoid infections? Our skin buddies actually do that for us. Some of the bacteria on our skin produce antimicrobial peptides that defend against pathogenic bacteria but are not harmful to your skin buddies. But when we use man-made antibacterial cleansers, we kill off the bad bugs as well as the good.

What if I told you that the death of our skin buddies

and damage to our gut lining are the two predominant causes of skin problems as we age—not sun exposure, genetics, or anything else you've been led to believe? Let's look at the evidence.

First of all, when it comes to sun exposure, your skin buddies can actually protect you against skin cancer despite your levels of exposure to the sun. There is a strain of bacteria that's common on healthy skin that can inhibit the growth of skin cancer. It does this by producing a compound called 6-N-hydroxyaminopurine (6-HAP), which kills several types of cancer cells but is not toxic to healthy cells. When mice with skin cancer received intravenous injections of 6-HAP every forty-eight hours over a two-week period, their tumor size was suppressed *by more than half.*[22] Pretty remarkable, isn't it? What if the current uptick in skin cancers (with more than a million new cases diagnosed in the United States each year) is being caused by the death of our skin buddies and the resulting lack of 6-HAP to protect us against skin cancer?

Furthermore, when patients with eczema, a painful skin condition that causes patches of skin to become red and inflamed, were treated with a type of bacteria that is normally present on the skin, they saw a significant reduction in symptoms.[23] Seeing these results, researchers began looking more closely at the skin flora of patients with eczema and found that they generally have a higher-than-normal population of bad bugs. In fact, an imbalance in skin flora is a sign of disease. Research shows that patients with primary immune deficiency conditions actually have altered skin flora. They tend to have more permissive skin, meaning that their skin flora includes species not normally found on healthy adults.

Just like your gut buddies, your skin buddies can trigger an immune response to a perceived threat. Normally, their ability to ring the alarm is helpful—think of any time you've gotten a cut and the skin around the injury became swollen. That's the result of your skin buddies inciting your immune system to attack any bad bugs that are trying to invade. Without such an immune system, bad bugs can take over the skin. But just like your gut buddies, sometimes your skin buddies overreact and create chronic inflammation—and skin conditions such as eczema can result.

Throughout your entire life, your internal and external skin function as the two front lines, where your gut and skin buddies fight off invaders (as long as you've been treating them right). When the bacteria in your gut are out of balance and allow bad bugs to permeate the gut barrier, your skin flora suffers, allowing bad bugs to take over. If your gut and skin buddies have left the building, you'll see it right there in the mirror—in the form of thinning skin, age spots, wrinkles, acne, and eczema. And if you're seeing those things in the mirror, it means the hordes at the gate have invaded and the war is on.

On the Longevity Paradox program, you'll nourish your skin buddies through the foods you eat and the products you use to care for the residents both inside and out. This is why my patients very often start to look younger after we've begun nourishing their gut buddies, even though we've made no changes to their skin care or cosmetics routine. This does not mean that skin care doesn't play a role in the program. But instead of products that kill off skin buddies and mimic hormones, you can make a few simple swaps that will support the health of your entire holobiome.

Here are some of the most beneficial ingredients, not for your skin itself but for your all-important skin buddies.

Bonicel (BC^{30})

We've known for a long time that BC^{30} is a beneficial spore-forming bacterium that, unlike most probiotics, is not digested by stomach acid. (Yes, that means that most of those other probiotics you've been taking probably haven't made it to your gut, but that's another story.) Ganeden, the probiotic company that patented

Bonicel, began applying BC^{30} to skin cells in a petri dish and found that the skin cells suddenly became fatter. The bacteria actually plumped up the skin cells. Ganeden developed a product called Bonicel, which is produced through a fermentation process and contains metabolites of BC^{30} yet is shelf stable for use in cosmetic products. In clinical trials, it has been proven to decrease coarse skin lines and skin shadows and increase skin hydration, skin smoothness, and skin elasticity.

I am such a big fan of Bonicel that I buy it to use in my own skin care formulations, but you don't have to get it from me. It is available in a wide range of antiaging skin creams, gels, masks, serums, body lotions, cosmetics, and hair care products.

Polyphenols

Polyphenols are important plant compounds found in many natural foods such as fruits, vegetables, grains (particularly lectin-free cereal grains, such as millet), tea, coffee, and wine. They give fruits and vegetables their vibrant colors and determine their flavor and aroma. Technically, polyphenols are characterized by the existence of more than one phenol unit—or building block—per molecule. That's what makes them "many"—or "poly"—phenols. But polyphenols aren't just the building blocks of plant-based chemicals; they also have a lot of incredible antioxidant properties with antiaging benefits.[24] In addition to stimulating autophagy and boosting cognitive performance,[25] they fight the free radicals that can enter your body every day if you're exposed to air pollution, cigarette smoke, or significant amounts of alcohol. But most polyphenols taken orally are absorbed poorly and not made into their bioactive compounds unless they are first eaten and transformed by the microbiome.

Your gut buddies absolutely love polyphenols. Resveratrol, which you read about earlier, is a potent polyphenol, and one of the reasons that olive oil is so good for your gut and brain health is its own high levels of polyphenols. But these compounds are equally beloved by your skin buddies. The polyphenols in pomegranates, for example, have a photo-protective effect on skin cells, meaning they can help your skin cope with molecular damage caused by sunlight.[26] One study showed that skin treated with pomegranate products experienced less collagen loss and skin protein breakdown and was better able to maintain its elasticity and youthful appearance.[27] Further studies have revealed that another type of polyphenol, ellagic acid, found in raspberries and blackberries, to name two sources, helped reduce hyperpigmentation (sunspots) on the skin when taken orally.[28]

One of the best sources of polyphenols for your skin is cranberry seed oil, the result of cold-pressing cranberries, which are filled with various polyphenols, each of which benefits your skin in its own unique way. The catechins in cranberries, for example, help fight signs of aging in the skin such as wrinkles and sagging by preventing cell stress and death. They also have anti-inflammatory and antibacterial properties.[29] Meanwhile, proanthocyanidins help protect the skin from the sun's harmful UVA and UVB rays[30] and help your skin buddies fight against viral skin infections.[31]

Other polyphenols found in cranberries include quercetin and myricetin. Quercetin, with its anti-inflammatory properties, helps soothe irritated skin,[32] while myricetin hydrates skin cells, keeps skin smooth and firm, prevents skin cell death, and combats damage from sun exposure.[33]

Of course, the anti-inflammatory actions of cranberries make them just as good for your gut as they are for your skin buddies. Each part of the cranberry, from its seeds to its juice, is packed with compounds that work to keep you healthy both inside and out. You don't have to wait until the festive season to embrace these little berries. Though they're seasonal, you can still reap their benefits through the many supplements and cold-pressed oils available on the market. When it comes to eating the berries themselves, it is always best to do so when they are in season and in moderation to avoid consuming too much sugar.

Polyphenol supplements are widely available, and many skin care brands are now including them in their products. Polyphenols are also present in large quantities in many of my favorite foods, including:

- Spices: Cloves, star anise, capers, curry powder, ginger, cumin, cinnamon, nutmeg
- Dried herbs: Peppermint, oregano, sage, rosemary, thyme, basil, lemon verbena, and more
- In-season dark fruits: Cherries, strawberries, cranberries, raspberries, blueberries, blackberries, huckleberries, pomegranates
- Natural beverages: Cocoa, green tea, black tea, red wine
- Seeds: Flaxseed (ground only), celery seeds, poppy seeds, black sesame seeds, *Nigella sativa* (black cumin or black caraway seed)
- Nuts: Chestnuts, pistachios, walnuts
- Wine, particularly red wine

- Oils: Extra-virgin olive oil, sesame oil, coconut oil
- Dark chocolate (70 percent cacao or more), cacao nibs (raw cocoa)

Wild Yam Extract

If you read the labels of your favorite skin care products and see *Dioscorea villosa*, know that it is simply the scientific term for wild yam. Yep, that's right. A potato's not-really-related cousin (wild yam is not to be confused with sweet potato) can help you achieve better-looking skin. Wild yams contain multiple compounds that can benefit several different parts of the body. They're filled with saponins, naturally occurring molecules in plants, that have anti-inflammatory, antimicrobial, and antioxidant properties.[34] In addition, they're able to dig easily into cell membranes, enabling them to work.[35]

The wild yam is best known for its high quantity of a compound known as diosgenin, a specific type of saponin. Diosgenin is used as an anti-inflammatory[36] and enhances DNA synthesis in human skin to restore skin cells.[37] Diosgenin is also believed to be effective as a skin depigmenting agent, helping protect you against nasty age spots.[38] It is used in cosmetics for its ability to fight the loss of collagen in skin.[39] Since collagen is the rebar in your skin, it plumps the skin, helping it maintain its youthful appearance and fresh look while holding everything together. Irritated and sensitive skin may reap benefits from the collagen-boosting effects of diosgenin.[40] Its anti-inflammatory and antioxidant properties come into play here as well, soothing skin while giving it the healthy nutrients it needs.[41]

Just like polyphenols, wild yam extract benefits both your skin buddies and your gut buddies with its powerful anti-inflammatory properties.[42,43] Internally, diosgenin helps stabilize your gastric and intestinal fluids. And just as it penetrates your skin cell membranes, diosgenin can penetrate your gut wall, which can help ease gastrointestinal discomfort stemming from inflammation more quickly.[44]

Wild yam is most commonly found as a supplement (dried as an herb in a capsule or tablet form) or as a 12 percent wild yam cream. It can also be bought as a liquid extract, which some people make into a tea. It's available in plenty of skin care products as well.

So now that you are up to speed on most of the residents that make up your holobiome, you're probably eager to find out exactly how you can begin taking better care of your tenants. That's exactly what we're going to cover next.

III

The Longevity Paradox Program

There is literally a life-and-death struggle constantly being waged between the good guys in your holobiome, which are dedicated to keeping you young until the day you (and they) die, and the bad guys, which desperately want to take over and run you into the ground. So the question is this: Which are you going to feed and nourish, and which are you going to starve out? This is a factor in aging that is *entirely within your control.* Every meal you eat and when you eat it, how and how much you exercise, the products you use in the shower, the supplements you take each day—all of these small choices add up to have a real impact on your life span and your health span.

For many, the conflicting advice out there—eat this, don't eat that; exercise this way, not that way—is frustrating and overwhelming. Many of us just throw our hands into the air and say, "Enough! I don't care how long I live; just give me what I need to feel good in the moment!" Well, I'm here to tell you that living for the present is not a winning strategy.

Recently I received a phone call from a good friend whose husband (let's call him Fred) has been a patient of mine for years. Fred was fit, a great businessman, and the life of the party. Fred's wife is a wonderful cook, and she and Fred regularly enjoyed many of the foods I advise against eating. For years, looking at his lab work, I warned him that trouble was brewing. Despite his being thin, his insulin levels were always high, which meant his brain was starving to death. His markers for damage to his arteries were always high, yet he easily passed the nuclear stress tests I performed on his heart.

Now Fred is in his early seventies and suddenly gave up tennis, a sport he loved, in the past six months because he began tripping on the court, falling, and bruising himself. Well, what the heck, he was getting old. Time to slow down, right? But in the last few months, his wife says, he has been searching for words and just sits in his chair all day long, watching TV. Last month, during his office visit, I detected that "faraway look" in his eyes and referred him to a local neurologist who specializes in dementia. So I was, sadly, not surprised when his wife called me to say that Fred had been diagnosed with Alzheimer's disease.

If Fred had changed course a few years ago, I would not be telling you about his current outcome. You have gotten this far in the book because you have decided that Fred's fate is not the fate you want. Let me assure you, you are not alone in this journey, and in the following pages you'll learn exactly how to make the best choices for your gut buddies—and therefore for you.

Chapter 8
The Longevity Paradox Foods

Imagine that you went to the hardware store, bought some grass seed, spread it out on your lawn, and then just left it there without any water or fertilizer. Obviously, the grass wouldn't grow. This may seem foolish, but many of my patients make the same mistake, but instead of their lawn, they do it with their microbiome. When they learn how important their gut buddies are to their health and longevity, they go out and buy expensive probiotics, swallow them, and expect their gut buddies to flourish and multiply. But if you don't fertilize those gut bugs with the foods they like to eat, they will quickly die off just like that grass seed.

To regrow a healthy, robust population of gut buddies, you must fertilize them! One of the many wonderful things about your holobiome is that its population of gut buddies can multiply rapidly. If you've read *The Plant Paradox*, you know that when you choose the foods your gut buddies like to eat, you can dramatically shift the bacterial population in your gut in a matter of days.[1] In this chapter, we'll cover exactly which foods your gut buddies love the most and which ones feed the bad guys. A full list of foods appears in the next chapter, but here we'll focus on the very favorites of both categories of gut bugs.

THE BEST FOODS FOR GUT BUDDIES

These are the foods that you should focus on eating as much as possible to ensure that your gut buddies will want to make your body their forever home.

Prebiotics

There is a lot of confusion about the difference between probiotics and prebiotics, but it's actually quite simple: whereas probiotics are the gut bugs themselves, prebiotics are the fibrous long-chain sugars they eat. Back to our gardening scenario: probiotics are the seeds you plant in your gut garden, and prebiotics water and fertilize them. They do this so well because they are indigestible by you. You can't digest them, so they stay in your gut, where your gut buddies can happily chow down on them. Remember our friends the naked mole rats and their steady diet of tubers, roots, and fungi? Well, tubers, roots, and fungi are wonderfully rich in prebiotics, which is why those mysterious creatures have such plentiful and diverse gut buddy populations that enable them to defy aging.

In addition to tubers such as yams, jicama, and tiger nuts, rutabagas, parsnips, sweet potatoes, mushrooms, taro root (cassava), yucca, celeriac, Jerusalem artichokes (sunchokes), chicory, radicchio, artichokes, and Belgian endive are all good sources of prebiotics, the last four also rich in our old friend *Akkermansia*'s favorite food: inulin. As a reminder, *Akkermansia* feeds on the protective mucus lining your gut wall and helps produce more of it. The more *Akkermansia* bacteria you have, the younger you will be into ripe old age. Here are a few more of my favorites.

Ground Flaxseed

Flaxseed has been around since 3000 BC, which means your gut buddies have had a long time to adapt to it. Remember, your gut buddies are creatures of habit. They prefer the foods they are most familiar with, and flaxseed definitely falls into that category. In fact, way back in the 700s, Charlemagne, the Holy Roman emperor, was so impressed by the health benefits of flax that he required his subjects to eat it. Of course, that was over 1,300 years ago, but he may have been onto something.

Flaxseed is beneficial because it contains not only prebiotic fiber but also a significant amount of lignans, a type of polyphenol. Flaxseed also has plenty of B vitamins and is one of the most significant sources of plant-based omega-3 fatty acids. Specifically, it contains a large amount of anti-inflammatory alpha-linolenic acid (ALA),[2] which supports the gut lining but, importantly, is not the same as docosahexaenoic acid (DHA), another omega-3 that you need for brain health. Many of my vegan patients try supplementing with flaxseeds to get their omega-3s, but humans simply cannot transform ALA into DHA. So flaxseeds are great for your gut lining, but you still need fish oil or algae DHA oil for your brain.

When flaxseeds are in their whole form, you cannot digest their beneficial compounds, so always choose ground flaxseeds, flaxseed meal, or flaxseed oil, which all have a slightly nutty taste and go virtually unnoticed in a smoothie or mixed into some coconut yogurt. But a word of caution: once ground, flaxseeds go rancid (oxidize) rapidly, so buy them whole and grind them in a coffee grinder, or buy the ground meal refrigerated. That goes for the oil as well: once opened, it will go rancid unless kept cool.

Because your skin buddies love flaxseed, too, you can use ground flaxseeds to make your own body scrub or use flaxseed oil to moisturize your skin and hair. I like combining flaxseed oil with various essential oils to create my own moisturizer that nourishes my skin buddies instead of killing them off.

Artichokes

Each one of these has more than 10 grams of prebiotic fiber. Not only that, but artichokes are also packed with vitamins: A, B, C, and E, as well as the minerals calcium, potassium, and magnesium, to name a few. Another benefit of adding the artichoke to your diet is its high antioxidant content and polyphenol content, which helps your liver. Plus they're delicious and fun to eat. You can prepare them whole and eat them that way, but if you want to make your life easier, I suggest buying frozen artichoke hearts, which brings the prep work down to nothing.

Leeks

These cousins of the onion are loaded with polyphenols and allicin, a compound that increases your blood vessels' flexibility and reduces cholesterol using a similar mechanism as statin drugs but without the side effects. Popular in Europe but not so much here, they are tasty and easy to prepare. Just make sure to slice them in half the long way and rinse them thoroughly or soak them in cold water before using—they have a lot of hidden dirt in between their layers. After that, use them in salads and soups, or grill them and then toss them into a salad. Delicious!

Okra

I know this is one of those foods that you either love or hate. For most people, it comes down to one thing: the texture. Sure, okra can be a little slimy, but it's a wonderful source of prebiotic fiber as well as vitamins C and A, iron, phosphorous, and zinc. In fact, fully half of its carbohydrates are prebiotic fiber. And I promise it's delicious when prepared correctly. My suggestion? Sauté it over very high heat, or roast it until it's crispy. And good news: if you can't find it fresh, don't worry; frozen okra is available in most grocery stores, and it's just as nutritious. Make sure to thaw it and pat it dry before cooking to cut down on the slime factor. Another fun fact: the mucus in okra binds to lectins!

Jicama

This delicious, slightly sweet, crispy vegetable tastes like a cross between an apple and a potato. And it's incredibly high in the prebiotic fiber inulin. Looking for more vitamin C in your diet? A 100-gram serving of jicama supplies 40 percent of your daily needs. Skip the orange juice, and eat some jicama sticks instead! You can cook with jicama—it actually stays nice and crisp in a stir-fry or sauté. But in my opinion, jicama shines when it's raw. It's good shredded into a slaw, chopped with cilantro and onion in a salsa, or—my favorite—cut into matchsticks or "chips" and used to scoop up guacamole.

Cruciferous Veggies

These vegetables, particularly broccoli, cauliflower, and Brussels sprouts, have profound gut buddy benefits. Brussels sprouts have tons of fiber as well as vitamins B1, B2, B6, C, and K. They are also rich in antioxidant and anti-inflammatory properties. Overall, Brussels sprouts are one of the most gut-friendly veggies out there. Meanwhile, broccoli has only slightly less fiber than Brussels sprouts, and a cup of cooked broccoli will give you as much vitamin C as a whole orange. It's also chock full of beta-carotene.

And did you know that broccoli contains vitamins B1, B2, B3, and B6? How much more B could you ask for? It's also a great source of iron, potassium, zinc, and magnesium.

Furthermore, there is evidence that broccoli can help heal a leaky gut. In one study, mice that had symptoms similar to those in humans who struggle with leaky gut and colitis saw a reduction in symptoms after eating a diet that included plenty of broccoli.[3] Mice with the same symptoms that did not eat broccoli weren't so lucky. Why did that happen? Cruciferous veggies, including broccoli and Brussels sprouts, contain a chemical compound known as indolocarbazole (ICZ) that binds with a receptor in the gut called the aryl hydrocarbon receptor (AHR), which is responsible for helping the body react to toxins. When AHR and ICZ bind together within the gut, the gut barrier and immune system are both strengthened.

Ironically, many of my patients with a leaky gut have previously been told by their doctors to avoid "roughage" such as broccoli and Brussels sprouts, but these vegetables can actually help heal their guts. If you suffer from a leaky gut or IBS, go slow and cook them to almost mush or use a pressure cooker when first introducing them to avoid diarrhea or cramps.

The Chicory Family

Though it's not a household word in the United States, chicory is in almost all of the salads that I am served in France and Italy. In fact, those clever long-lived Italians have a *tricolore* salad consisting of Belgian endive, radicchio, and arugula. The first two are forms of chicory, while the latter is a cruciferous veggie. A gut buddy's dream food, the inulin in chicory feeds *Akkermansia.* Try all the forms available, such as escarole, curly endive, radicchio, Belgian endive, and more.

Nuts

You read earlier that certain nuts nourish the gut buddies that produce butyrate to support your gut lining and your mitochondria. But when it comes to nuts, you've got to be careful. It's a little-known fact that some "nuts" are really just seeds. Take cashews, for example: they are seeds, and they are full of lectins. Other "nuts" such as peanuts are actually legumes. And legumes are lectin bombs. It's best to stay away from peanuts, even if you're not allergic to them.

But real nuts can do wonderful things for your gut buddies and as a result support heart health, reduce your chances of developing gallstones, help protect against diabetes, regulate blood pressure levels, and protect against inflammation. The nuts your gut buddies like best are walnuts, macadamia nuts, hazelnuts, and pistachios. Peeled almonds and blanched almond flour are fine, but many of my autoimmune patients react to the brown peel on almonds. I recommend eating a handful of nuts every day. Your gut buddies will thank you.

Mushrooms

Mushrooms have long been touted for their health benefits, but for the first time, my fellow longevity researchers have pinpointed two important antiaging compounds—ergothioneine and glutathione—that are plentiful in mushrooms. Mushrooms are the highest dietary source of these two antioxidants together, so they protect you from free radicals and help you stay young.[4]

Of all the mushrooms tested, porcinis had the highest levels of polyphenols by far. (Is it a coincidence that this species of mushrooms is popular in Italy, home of some of the world's longest-lived people?) The second runner-up was the popular white button mushroom. Mushrooms contain huge amounts of polysaccharides, which feed your gut buddies, which, in turn, tell your immune system to "chill out." It's this feature that gives them their ability to enhance the immune system. Best of all, unlike some foods that lose their nutritional value after cooking, exposure to heat doesn't seem to affect the key polyphenols in mushrooms. So throw them into stir-fries or sauté them by themselves as a delicious anti-aging vegetable side dish.

But the real reason to get more "'shrooms" into your diet is their high content of polyamines, the longevity-promoting compounds that are found in high levels among centenarians! My favorite polyamine is spermidine, which was named for its presence in semen. (I kid you not.) Studies show that spermidine increases life span and is cardioprotective.[5] Mushrooms are one of the great sources of spermidine! You can imagine another great source. (Hold the giggles, please.)

Low-Sugar Fruits

Though most fruits should be eaten in moderation and only in season (more on this in a moment), some fruits are naturally low in sugar and can be eaten in large quantities year-round. Sadly, we often don't recognize these antiaging fruits as fruits at all. They include the following.

Avocados

Yes, this guacamole star is actually a fruit! And its flesh contains even more prebiotic fiber than an artichoke.

Like the artichoke, the avocado boasts significant amounts of vitamins C and E and potassium, as well as a lot of folate. Avocados are also chock full of the same healthy monounsaturated fat found in olive oil, oleic acid, which supports brain function and is important to consume at any age.

Your skin buddies also adore avocados. Try mashing a ripe avocado and using it as a DIY skin mask or a deep conditioner for your hair. The fatty acids in avocados can help your skin's natural oil barrier stay strong, protecting you from the aging effects of sun damage.

Green Bananas

Everybody praises bananas because they're rich in potassium, but they're also rich in something else that your gut buddies definitely don't love: sugar. One medium banana contains 14 grams of sugar—that's 3½ teaspoons! But unripe tropical fruit has lower fructose content and is made up of resistant starches, which, as you read earlier, your gut buddies love to munch on. So if you can find green bananas, go for them. Like avocados, these low-sugar fruits nourish your hair and skin. Try mashing half an avocado and half a green banana together and using the mixture as a hair and/or skin mask. You'll smell great, and your skin buddies will love it.

Raspberries, Blackberries, and Mulberries

These tart berries have only about 5 grams of sugar per cup, are an awesome source of prebiotic fiber, and are chock full of polyphenols including ellagic acid. They also have all the vitamins A, C, and K you could hope for. So if you have a sweet tooth, try freezing some raspberries and noshing on them when you need to satisfy a craving without supporting the bad bugs in your gut.

Figs and Coconuts

Believe it or not, neither of these popular fruits is actually a fruit. Figs are flowers, not fruits, and coconuts are technically tree nuts! Along with the pomegranate, the fig is probably the oldest known "fruit," dating back in records to 5000 BC. As the cover of this book suggests, the fig is ripest and at its best when it is in full flower. Never seen a fig's flower? Just cut the fig in half; the insides are actually multiple tiny flowers waiting for a specific wasp to crawl through the little hole in the base and complete the fertilization process. So a ripe fig is at a very old age and ready to make new life—hence earning its rightful spot on the cover of this book. Good news: most of the sugar in figs comes from prebiotic fiber. And when they're ripe (between August and December), they're especially delicious.

Coconut is one of my all-time favorite foods—not

only because it tastes so good but also because it has about 7 grams of prebiotic fiber per cup. You don't have to eat just the meat—you can bake with coconut flour or sprinkle shredded coconut on top of grilled veggies. Just make sure you're using unsweetened fig and coconut; a lot of dried fruits are loaded with sugar. Check labels carefully. And please ditch the "healthy" coconut water: it's pure sugar water!

Healthy Fats

The type of fat you eat is very important because when it comes to inflammation, most fat sources aren't neutral—they are either proinflammatory or anti-inflammatory. But they aren't that way by nature. For instance, omega-3 fats from fish oil are anti-inflammatory, right? Well, not so fast. It turns out that the real anti-inflammatory compounds made from DHA and EPA (two types of omega-3s) in fish oil are called resolvins,[6,7] and these guys are the superheroes of blocking inflammation in your nerves and eyes. But here's the caveat: you need a little bit of the active ingredient in aspirin (salicylic acid) to get these effects. That's why I recommend taking an 81-milligram enteric-coated aspirin a few times a week to activate that fish oil you've been swallowing.

And how about that evil omega-6 fat arachidonic

acid (AA), the supposed cause of inflammation? Well, here's a paradox again. Half of your brain's fat is the omega-3 fat DHA, while the other half is AA! What's that stuff doing up there? It's actually preventing inflammation in your brain and its memory center, the hippocampus.[8]

What's more, in a large study of Japanese men published in March 2018, the men with the highest levels of AA and another omega-6 fat, linoleic acid (LA), had the lowest risk of death from all causes and the lowest risk of cardiovascular deaths![9] And in athletic performance trials performed at Baylor University, athletes who supplemented with AA not only improved their performance compared with those who took a placebo, but (spoiler alert) a marker for inflammation that I follow, interleukin 16 (IL-16), went down significantly, not up![10] What a big, fat paradox. That evil arachidonic acid turns out to be a sweetheart!

The omega-3s EPA and DHA do far more than quell inflammation. Several studies show that humans with the highest levels on the omega-3 index (the amount of EPA and DHA in your blood as measured over the previous two months) have the largest brain size and the largest memory areas, the hippocampus, compared with those with the lowest levels.[11] And, as you read earlier, my vegan patients generally don't know that flaxseed oil, with its short-chain omega-3 fat, doesn't convert to EPA and DHA. When I first see them, they normally have horribly low omega-3 indexes unless they are taking algae-derived DHA. How important is this? Consider the findings of a study from Oxford University that looked at the learning ability of students supplemented with algae-derived DHA or a placebo. Those who took the DHA exhibited improved learning and behavior, and students afflicted with ADHD saw an improvement in their symptoms as well.[12] Omega-3 supplements have also been shown to reduce disruptive behavior in healthy children.[13]

So long live fish oil, algae-derived DHA, and arachidonic acid! And where can you get both long-chain omega-3s and omega-6s? Shellfish are probably the best choice, while egg yolk contains plenty of arachidonic acid alone. But there are plenty more interesting fats and oils. Here are some of my favorites.

Perilla Seed Oil

This little-known oil is derived from the perilla plant, which is in the same family as mint and basil. For centuries, people in China have used perilla seed oil to help ease colds and coughs and prevent the flu. And it's the most popular oil in Korea. But it's also wonderful when it comes to maintaining a healthy body as you age, as it supports both joint and heart health. It's a pretty significant plant-based source of omega-3 fatty acids, making it a good option for vegetarians who need to increase their omega-3 intake. Like flaxseeds, it's particularly rich in the omega-3 alpha-linolenic acid (ALA), which benefits your cardiovascular system, as well as rosmarinic acid, which has antibacterial, antiviral, antioxidant, and anti-inflammatory properties. As you will learn, this may be the secret of brain health for the folks in Acciaroli, Italy. Perilla oil works great as the fat in stir-fries, eggs, or Miracle Noodle brand pasta (made of konjac plant fiber) or even used in a salad dressing. And for a super aromatic twist on pesto, try including it in your recipe, combined with olive oil.

MCT Oil

MCT stands for "medium-chain triglycerides"; MCT oil can be turned into ketones by your liver. As you read earlier, ketones are normally enzymatically manufactured from stored fat once your sugar supplies start running low. So at night when you're not eating, the cells in your brain and the rest of your body make use of the energy from ketones to fuel mitochondria. MCT oil is sometimes called liquid coconut oil because it remains liquid even at cold temperatures, unlike coconut oil, which becomes a solid. For the most part, MCTs are burned as fuel, instead of being converted to fat.

Olive Oil

As you read earlier, olive oil is a delivery system for polyphenols and therefore is a miracle drug for longevity. I strive to consume a liter a week as people in so many of the Blue Zones do, and I encourage you to do the same! Moreover, olive oil has been shown to increase a type of cholesterol called apolipoprotein A-IV, which is fundamental in preventing platelets from clumping after eating a meal or in the early morning, when heart attacks are more likely to occur.

Other good fat sources include:

- Macadamia nut oil
- Walnut oil
- Avocado oil
- Thrive algae oil, a great cooking oil
- Citrus-flavored cod-liver oil
- Coconut oil. Unless you carry the APOE4 genotype, don't buy into the warnings about coconut

oil. The Kitavans, whose diet consists of 30 percent coconut oil and who don't have heart disease or strokes, apparently haven't gotten the message about how dangerous it is.

Dairy Alternatives

As you know, most conventional dairy products are highly inflammatory because they contain casein A1, a type of dairy protein that stimulates inflammation. Luckily, there are plenty of options that are far better for your gut buddies and are just as satisfying for you.

Goat Cheese/Yogurt/Butter

Let me be clear: there is no need for human beings to drink milk from another animal to support or benefit health. You or your child is not a baby cow, kid (baby goat), lamb, or water buffalo. These milks are designed to make the respective baby animal grow quickly to lessen the chance of predation. All contain impressive amounts of insulin-like growth factor 1 (IGF-1), which, as you learned previously, promotes not only growth but also cancer and aging. And all contain milk sugars such as lactose, which, after all, is a sugar. By now you know what they do.

Goat or sheep cheese, though made from goat or

sheep milk, lacks IGF-1, a water-soluble portion of milk, as well as the sugars, which are discarded or used up in the process of fermentation. If you want animal protein in your cheeses, these are your best options. On the other hand, coconut yogurt lacks animal protein as well as IGF-1 and may be your best yogurt choice.

Coconut Milk and Yogurt

All right, coconut milk isn't really milk. And it's not the water that pours out of the coconut when you crack it open, either, as many people think. That stuff is pure sugar, folks! Rather, coconut milk is what you get when you blend and strain the meat of the coconut. Among the many benefits of coconut milk is the fact that it's full of a really good-for-you fat called lauric acid. I personally love the natural sweetness and creamy richness of coconut milk. There are also great coconut yogurts on the market that are worth a try. But please use only unsweetened coconut milk or yogurt. If you see sugar on the label of unsweetened coconut yogurt, not to worry. Since coconut meat has virtually no sugar, sugar is added to make coconut yogurt because the bacteria that make yogurt have to have sugars to ferment. The bacteria eat the added sugars, so they are all gone by the time you have the finished product in your hand.

Ghee

Ghee is clarified butter, meaning that the butter has been melted so the proteins and fats separate by density. Once the butter is simmered, any impurities are skimmed off the top. Next, the liquid fat is retained, while the solid residue, which consists of the casein and a little bit of whey protein that settles on the bottom, is discarded (including any casein A1, which means you can safely eat ghee even from casein A1 cows). Ghee is a staple of South Asian and Indian cuisines, as it can be stored without refrigeration and won't go rancid.

Millet Is Not Just for the Birds

Though most grains are inflammatory and aging thanks to their high levels of lectins, millet is a notable exception. It's the key seed in birdseed, but it's not just for the birds. Millet is a kind of seeded grass, and its many variations are actually grown all over the globe. Millet is lectin free and has been getting a lot of attention lately because those with celiac disease have been gravitating toward it when looking for gluten-free options. It is also rich in magnesium and potassium and high in fiber. But the best thing about millet is that it can work wonders for your gut. Your gut buddies love its fiber because it allows them to better digest the foods you eat and retain their nutrients. As you well know if you've been reading carefully, you are not what you eat; you are what your gut buddies digest—and millet helps them do exactly that!

Coffee Fruit

I have some very good news: your daily coffee habit may actually help you live longer. In addition to pepping you up, coffee consumption is associated with a reduced risk of death. Not a fan of coffee? No problem—you can get the same benefit from coffee fruit, which is the fruit from which coffee beans are pulled (like cherry pits from a cherry). Believe it or not, this fruit grows on a flowering, bushy green plant. When the fruit is ready to be picked, it is a rich burgundy color (though some less common varieties are actually yellow or green). The fruit of the coffee bean—known simply as coffee fruit—has a taut skin, but once you break through it, you find a sweet, delicious, gummy pulp with a sort of herby or vaguely floral melon taste. It's light and delicious and can keep you young in important ways.

For starters, coffee fruit is packed with antioxidants and polyphenols that help boost your immune system, assist in protecting your body against free-radical damage, and fight inflammation. But it also has loads of flavonoids, plant chemicals that are potent antioxidants and anti-inflammatories in their own right. Flavonoids also promote the production of nitric oxide, a natural gas produced in your body that communicates between cells. In particular, nitric acid tells your blood vessels to relax (dilate) so that blood can flow through them more freely. Of course, as we age and inflammation takes its toll, our blood vessels tend to constrict more and more. So the flavonoids and nitric acid in coffee fruit help keep your blood vessels flexible and young.

Coffee fruit also supports better cognitive function by boosting brain-derived neurotrophic factor (BDNF), which, as you read earlier, helps your brain grow new neurons.[14] Maybe that's why you feel so much sharper and more alert after drinking a cup of coffee! Unfortunately, raw coffee fruit is incredibly hard to find unless you live on a coffee farm. But the dried form is easy to find in supplements. Can't find coffee fruit? Then stick with coffee, which has most of the same benefits.

Extra-Dark Chocolate

Who doesn't love chocolate? Not your gut buddies, that's for sure! They want you to eat chocolate every single day, so go ahead and indulge in an ounce of extra-dark chocolate. Not only will you get to eat something satisfying and decadent, but you'll also be doing something great for your health. Chocolate contains antioxidants and flavonoids, both of which have powerful anti-inflammatory properties. But the real benefits lie in plant-derived cacao, which is the main ingredient in most commercial chocolate.

The flavonols in cacao can provide a boost to brain health, and over time, consuming flavonol-rich chocolate can protect the brain. In one study, people who ate small amounts of dark chocolate for a period of three months had better memory, processing speed, and attentiveness than did a control group. Other research has shown a notable improvement in memory retention and new learning in elderly adults after consuming these flavonols. This is believed to be tied to increased blood flow to the cerebral cortex, the brain structure that is most affected by aging. Simply put, the part of the brain that deteriorates the most with aging gets more blood flow when you eat chocolate, preventing this deterioration from happening.[15] Pretty sweet (pun intended), right?

But before you grab that candy bar in the checkout line, remember, milk chocolate has no health benefits—it's mostly sugar. Look instead for chocolate that's at least 72% cacao. The darker the chocolate, the better. And it doesn't take much to do the job. According to the European Food Safety Authority (EFSA), the recommended daily amount of cocoa consumption is 200 milligrams (about .007 ounce) for maximum benefits. So keep your chocolate consumption to about an ounce a day (a couple of small squares in a standard bar) for maximum health benefits. I personally prefer 90% or greater chocolate. And avoid dutched or alkali chocolate products; the dutching process destroys the health-promoting polyphenols in chocolate.

Green Tea

My favorite hot beverage, green tea, improves the symptoms and reduces the pathology of autoimmune disease. It does this by suppressing the proliferation of autoimmune T cells (the cops in your inner condo) and their inflammatory cytokines.[16] I drink about five cups of green and mint teas a day and recommend that you do the same to keep inflammation at bay whether you are suffering from an autoimmune condition or not.

For an extra longevity kick, try organic pu-erh tea. This fermented tea has been shown to reduce the oxidation of lipids.[17] Its polyphenols also help lower iron levels. No, you won't get "iron-poor blood." As you read earlier, iron is one of the signature aging-promoting compounds.[18] We literally rust out if our iron level is on the high side.[19] And you already know that blood donors live longer than nondonors do. Last, pu-erh tea promotes the growth of *Akkermansia muciniphila*, and who doesn't want more of them around?

So we've got coffee, tea, and chocolate, as well as delicious foods such as avocado, olive oil, coconut, tubers, mushrooms, raspberries, blackberries, figs, and pomegranates. And you thought your gut buddies didn't have good taste in food! It turns out that it is remarkably easy to feed your gut buddies while still enjoying your meals. Just make sure not to inadvertently feed the bad guys along the way.

GUT-DESTROYING BAD BUG FAVORITES

Though I generally like to avoid speaking in absolutes, the foods that follow are the main sources of nutrition for the bad bugs in your gut and should be avoided as much as possible. If you slip up and consume any of them, it's okay; just refocus on feeding your gut buddies so they'll proliferate and drive out the bad bugs. Then come back to this list for a reminder of which foods are your bad bugs' favorites and therefore the worst ones for your longevity.

Simple Sugars and Starches

My apologies, but then again, you probably already knew this was coming. Whether it's glucose, fructose, or sucrose, any type of simple sugar is the number one food of choice for bad bugs everywhere. And yes, that includes the sugar in fruit (which is fructose). As you now know, we were never meant to consume fruit year-round. Before globalization, sweet flavors were readily available only during the summer and fall, when humans needed to store fat by gaining weight by eating fruits in preparation for the winter period of regression. But now we are living in an endless summer, and fruit, sweet treats, and real or fake sugars are available around the clock. This is a driving factor of the obesity epidemic, which, of course, if you've been following along so far, is really being driven by bad gut bugs.

The bad bugs love sugar, and remember, so do cancer cells. So as painful as it may be to begin with, cutting down on sugar is the best thing you can do to drive out bad bugs and help your gut buddies win the war. In addition to table sugar, sweets, and other obvious forms of sugar, avoid the following fruits that are highest in sugar, especially when they are not in season.

Grapes

Grapes are an easy snack, and I loved them as a kid. In fact, most kids love them. Do you know why? Because they're essentially tiny sugar bombs. A cup of fresh grapes has around 23 grams of sugar. That's 6 teaspoons of sugar. It's dessert, not a healthy snack! However, when fermented into wine or vinegar, grapes are amazing for you. They're a high-polyphenol food, and the fermentation process removes the sugar, making them much safer. So enjoy plenty of balsamic vinegar and moderate amounts of red wine—just skip whole grapes! But let me issue my famous proviso again: if you don't already drink alcohol, don't start!

Mangoes

Of course, a mango can vary in size, but according to the USDA, an average mango has up to a whopping 46 grams of sugar (that's 12 teaspoons). Mangoes are full of all three kinds of sugar: glucose, fructose, and even sucrose. And as a mango ripens, all three types of sugar increase. This is why fresh mangoes are so delicious and so prized by the bad bugs in your gut. But unripe mangoes are pure heaven to your gut buddies, as they are pure oligosaccharides. I eat unripe mango salads regularly, and they are delicious.

Ripe Bananas

Before a banana ripens, it is made mostly of resistant starch. In fact, as you read earlier, green bananas are made up of almost 80 percent resistant starch. But once the banana ripens, this starch gets converted into sugars—so much sugar that a "serving size" is actually half of a large banana. Who eats just half a banana? And you'll find sucrose, fructose, and glucose in a ripe banana. So stay away and opt for unripe green bananas instead. I promise that not only will you get used to them, but your gut bugs will thank you for the gift that keeps on giving.

Lychees

These little guys have a sweet, floral scent and a tart taste. They're sometimes served with Asian-inspired meals. They'll fool you because they don't seem too sweet, but they're chock full of sugar—about 29 grams, or 7 teaspoons, per cup. If you haven't tried lychees yet, don't start now.

Apples

You've been told since you were a kid that "an apple a day keeps the doctor away." But did you know that one medium apple contains 19 grams (5 teaspoons) of sugar? So the saying really should be "An apple a day keeps the gut buddies away." Now, apples are also really high in soluble fiber, so they're not a complete no-go. But stick to in-season (that means August through November) apples, and think of them as a special treat, not a daily snack.

Pineapple

I've always felt that pineapples taste too sugary. And with 16 grams (4 teaspoons) of sugar in every cup, you can bet that I avoid them at all costs. You should, too.

Pears

Finally, a medium-ripe pear contains about 17 grams of sugar. But, great news, crispy pears such as Anjou and unripe Bartlett are full of resistant starches, so please enjoy them! And when that fancy box of pears arrives for the holidays, eat them just before they ripen, and consider them a gift for your gut buddies.

Sugar Substitutes

As you read earlier, sugar substitutes such as sucralose, saccharin, and aspartame are just as bad for your gut health as actual sugar, if not worse. They alter the gut microbiome, encouraging bad bugs to take over![20] And a recent study showed that consuming sucralose, known as Splenda, promoted higher blood glucose and insulin levels on a glucose tolerance test in humans than did drinking water.[21] Moreover, despite what the government has told you, sucralose is not inert and is converted into toxic compounds that may persist in you for weeks.[22]

Artificial sweeteners also promote weight gain. This is because when a sweet substance attaches to your tongue, your receptors taste sweetness. Then your tongue's nerves spring into action. They tell the pleasure receptors in your brain's reward center to get more of this amazing sugary food. Why? Because you'll need it to store fat in anticipation when the season changes and there isn't any food around.

Artificial sugar is designed to send the same pleasure signal to your brain as real sugar does. But when the calories from real sugar don't make it to your bloodstream (because you never ingested glucose to begin with), your brain feels cheated and gets mad that the sugar it was promised never arrives. So what does it do? It tells your body to go back and get more sugar. That's why I was addicted to eight Diet Cokes a day while still being obese! Stop frustrating your brain and your gut buddies by consuming artificial sugars!

Conventional Dairy Products

Got milk? I hope not. You already know about the dangers of casein A1, which can spark autoimmune attacks. These reactions, of course, get worse the more cow milk, cow cheese, and cow ice cream you ingest. In fact, most people who complain of lactose intolerance—and all the pain, discomfort, and embarrassing symptoms that come with it—are actually struggling with casein A1 intolerance.

But don't be dismayed. There are a number of herds of cows (as well as goats, sheep, and buffalos) that make a different protein—casein A2—which is much better for you. When it comes to dairy products, it's simply a matter of choosing the right type of protein from the right type of animal. I've been delighted to see that since the publication of *The Plant Paradox*, casein A2 products have become widely available. Look out for them and go back to enjoying certain dairy products without destroying your health.

Casein A2 is present in the milk of goat, sheep, and water buffalo and is found in imported cheeses from France, Italy, and Switzerland. There are even goat, sheep, and water buffalo butters, which have a translucent white color because goats, sheep, and buffalos transform the copper-colored beta-carotene in what they eat into colorless vitamin A, while cows skip that step.

Have you ever tasted buffalo mozzarella, otherwise known as *mozzarella di bufala*? While the vast majority comes from the Naples region of Italy, a new manufacturer makes it from the milk of grass-fed water buffalo in Uruguay, imported to the United States as Buf. Look for it in Whole Foods, or visit its website to find a local merchant. Buffalo mozzarella has a creaminess that is unparalleled.

Bad Fats

Though fat in and of itself is not bad, there are certain fat sources that should be avoided for the sake of your longevity. They include the following.

Saturated Fats

Many of my friends in the paleo and ketogenic communities laud the health benefits of saturated fats. But they are sadly overlooking a major problem with these fats: those sneaky LPSs, the fragments of bacteria that are constantly being produced as bacteria divide and die in your gut, travel through your gut wall and out into the body by riding on and hiding in saturated fats. Then they are transported directly to the hunger center in your brain, the hypothalamus. There, the resulting inflammation sparks hunger. This is why folks on the paleo diet are often sidelined by hunger. So say good-bye to saturated fats and the LPSs that come with them!

Peanut Oil

When my colleagues at the American Heart Association looked at the effects of different types of fat on the health of arteries (which are critical to healthy heart function), they found that peanut oil led to the most widespread and advanced atherosclerosis and the most severe coronary narrowing. Of course, we know that this is because peanut oil is full of lectins, which leads to the autoimmune attack on the arteries.[23]

For a full list of fats to avoid, see pages 318–20. But for now, here is a short list of the most damaging fats that cause inflammation and allow bad bugs to take over.

- Grape seed oil
- Corn oil
- Cottonseed oil
- Safflower oil
- Sunflower oil
- Partially hydrogenated vegetable or canola oil

So in the end, the news isn't that bad after all. You can avoid simple sugars, overly sweet fruits, artificial sweeteners, conventional dairy products, and bad fats for the sake of your gut buddies, can't you? Just a few more tweaks to your daily habits, and your gut buddies will be relaxing in the comfort of their newly renovated luxury suite.

Chapter 9
The Longevity Paradox Meal Plan

As you read earlier, I believe that one of the best books on longevity ever written was published in the 1500s. At the age of 40, the author, Luigi Cornaro, was in poor health, which his physicians attributed to his excessive eating and drinking. He began to adhere to a self-designed diet that we would now describe as calorie restricted, which he maintained well into his old age. In fact, on one occasion when his friends and family insisted that he increase his calorie intake, he described in vivid detail how much worse he felt on that regimen. He then returned to his usual style of eating. Ultimately he died a young man at the ripe old age of 102.

Cornaro began writing *How to Live 100 Years, or Discourses on the Sober Life* (subtitled *The Sure and*

Certain Method of Attaining a Long and Healthful Life) when he was in his eighties and added a new chapter as he reached each five- or ten-year milestone thereafter. In his book, he rejected the conventional wisdom that we are meant to spend our later years in a state of decline. He wrote:

> *And now some sensual and unreasonable individuals pretend that the existence of a man after he passes the age of 65 cannot be termed a living one, but a dead one. I will plainly demonstrate that they are mistaken, for I have a desire that all men should attain my age, which is the most beautiful period of life.*

I couldn't have said it better myself. But don't take my word or his for it. A 2018 study comparing adults aged 60 to 90 to those aged 18 to 36 showed that the feeling of being in control of their lives enabled the older group to feel and act similarly to the younger group.[1] This is what I want for you as well: to attain old age and enjoy it as the most beautiful period of your life. In our culture we have a tradition of saying that "youth is wasted on the young"—meaning that the physical vitality of youth is wasted when we have less wisdom and experience to direct our energy and enjoy our lives. But, as I hope you have seen by now, it doesn't have to be that way.

Now, it's no coincidence that Luigi Cornaro reached this robust state of health after dramatically cutting his food intake. On the Longevity Paradox program, you, too, will plan periods of calorie restriction, but you'll do it intermittently and without suffering. Let's take it one month at a time. Each month, your days will be broken down as follows:

FAST-MIMICKING DAYS Five consecutive days a month, you will eliminate animal protein and limit calories to 900 a day to mimic the benefits of a whole month of full-time calorie restriction.

FREE DAYS On most days, you will eat as much as you like of your gut buddies' favorite Longevity Paradox foods.

BRAIN-WASH DAYS Once or twice a week, you will skip dinner or eat it very early to make sure your brain is scrubbed completely clean while you sleep.

OPTIONAL CALORIE RESTRICTION DAYS If you choose, you will consume only 600 calories a day once or twice a week to get extra longevity benefits. This can even be done on brain-wash days for a double benefit.

OPTIONAL INTENSIVE CARE CLEANSE If you are suffering from degenerative problems or want to kick-start the process, this program, which includes more fasting and brain-wash days, will give your mitochondrial function an extra boost.

Now let's break down each of these components and take a detailed look at how and what you'll be eating all month long.

FAST-MIMICKING DAYS

You now know how important calorie restriction is for your health and longevity. The great news is that you can restrict calories for only five consecutive days out of the month and still reap the benefits of an entire month of calorie restriction. That's right. My friend and colleague Valter Longo, head of the Davis School of Gerontology at the University of Southern California, has shown that a monthly five-day modified vegan fast gives you the same longevity-boosting results as a month of a traditional calorie-restricted diet does.[2]

I strongly recommend that you begin the Longevity Paradox program by doing five fast-mimicking days in a row. Not only will you get the same benefits as if you had restricted your calorie intake for the whole month, but you will dramatically change the makeup of your gut bacteria in those five days, driving out the bad bugs and nourishing your gut buddies. In fact, research by Dr. Longo and by researchers at the University of Colorado, Boulder, confirms that a fast-mimicking diet changes the types of bacteria in your gut for the better.[3] In a study conducted by researchers at the Knight Lab at the University of Colorado, participants who followed just a three-day cleanse saw dramatic shifts in their microbiome, including increased levels of our good friend *Akkermansia*! And once your gut buddies are in good shape, they will make it much easier for you to follow the rest of the program.

Please don't worry that you will starve to death by cutting calories for a few days. Humans can easily go without food for two months and longer if water is available.[4] If you are overweight or obese, have insulin resistance or an elevated fasting insulin level, or are on insulin injections, I would advise you to check out the advice in chapter 10 of *The Plant Paradox*, which will guide you in making the transition from insulin resistance to insulin sensitivity and help you start using ketones as fuel more quickly and easily. If you do find yourself getting hungry and really needing something more, feel free to have about a tablespoon of MCT oil up to three times a day to ward off the "low-carb flu." As you grow more insulin sensitive and your gut buddies take over and start renovating, you will be able to complete these five days easily without hunger becoming an issue.

So what will you be eating during these five days? Actually, the things you *won't* be eating are far more important, so let's start there.

Foods to Avoid

All dairy products
All grains and pseudograins
All fruit, including all seeded vegetables, which are technically fruits
All sugar sources
Unapproved seeds
Eggs
Soy products
Nightshade plants (eggplant, peppers, tomatoes, potatoes)
Corn, soy, canola, and other vegetable oils
Meat, chicken, and all other animal products

All other foods on the "Gut-Destroying Bad Bug Favorites" list (page 288)

Foods to Include

And what can you eat? Your gut buddies' favorites, of course, which include the following.

Vegetables

You can eat as much as you'd like of all the following vegetables, either cooked or raw. If you have irritable bowel syndrome, SIBO, diarrhea, or another gut issue, limit your consumption of raw veggies and cook the rest of the things you eat thoroughly. All vegetables should be organic and can be purchased either fresh or frozen. If fresh, they should be in season and grown locally with sustainable farming practices, if at all possible.

- Cruciferous vegetables: Bok choy, broccoli, Brussels sprouts, Swiss chard, any color and type of cabbage, cauliflower, kale, mustard greens, collard greens, rapini, kohlrabi, watercress, mizuna, arugula
- Greens of all kinds: Belgian endive, all kinds of lettuce, spinach, dandelion greens, chicory

- Treviso, radicchio
- Artichokes
- Asparagus
- Celery
- Fennel
- Radishes and other root vegetables such as yams, taro root, jicama, yucca, cassava, turnips, rutabagas, horseradish
- Fresh herbs: Mint, parsley, sage, basil, and cilantro, plus garlic and all kinds of onions, including leeks and chives
- Ocean vegetables: Kelp and seaweed, including sheets of nori

Protein

For these five days, you are going to go vegan. That means no eggs, meat, chicken, or dairy products of any kind. Do not worry that you will become protein deficient! Remember, you are probably eating too much protein right now, and your body recycles the protein that is already present. Eliminating animal products for five days gives your body a rest from digesting all that protein and allows it to become an eco-friendly resort for your gut buddies! Sources of plant-based protein that you can eat during these five days (in quantities of eight ounces a day or less) include but do not have to include:

- Tempeh (fermented soy, without grains)
- Hemp tofu and hemp seeds
- Pressure-cooked legumes such as lentils and beans
- Hilary's Millet Cakes
- Approved nuts and seeds

Remember, your great-ape cousins and your ancestors got plenty of protein by eating leaves, and you can as well.

Fats and Oils

Acceptable vegetable fat sources for these five days include:

- Avocado—feel free to have a whole one each day
- First-cold-pressed extra-virgin olive oil

- Olives of any kind
- Nuts: Walnuts, macadamia nuts, pistachios, hazelnuts, pine nuts, Marcona almonds, blanched almond flour
- Avocado oil
- Coconut oil
- Macadamia nut oil
- MCT oil
- Perilla oil
- Sesame seed oil
- Walnut oil
- Hemp seed oil
- Flaxseed oil

Condiments and Seasonings

Because of their sugar content (not to mention other harmful ingredients), avoid all commercially prepared salad dressings and sauces. Instead, use as much as you like of the following.

- Fresh lemon juice
- Vinegars
- Mustard
- Freshly ground black pepper
- Iodized sea salt
- Your favorite herbs and spices, minus red chili pepper flakes

Beverages

Obviously, you will avoid all sodas (including diet soda), sports drinks, lemonade, and other commercially prepared beverages. Instead, enjoy at least eight cups of tap or filtered water a day, as well as:

- San Pellegrino or other Italian sparkling mineral water (or Acqua Panna, a still mineral water)
- As much tea as you'd like—green, black, or herbal
- Regular and/or decaffeinated coffee (black or with unsweetened almond, hemp, or coconut milk)

- Stevia extract (preferably SweetLeaf), Just Like Sugar (inulin), or monk fruit to sweeten your tea or coffee, if you like

I put this program together with my good friend Irina Skoeries, who also did the Three-Day Kick-Start Cleanse of *The Plant Paradox*, with rigorous standards to make sure you will get the right number of calories and amount of protein on each day of the "fast." Using the following meal plan, you will duplicate the effects of a month of calorie restriction while stimulating stem cell regeneration and strengthening your gut wall.

DAY 1

BREAKFAST	Green Smoothie
SNACK	Romaine Lettuce Boats Filled with Guacamole
LUNCH	Arugula Salad with Hemp Tofu, Grain-Free Tempeh, or Cauliflower "Steak" and Lemon Vinaigrette
SNACK	Romaine Lettuce Boats Filled with Guacamole
DINNER	Cabbage-Kale Sauté with Grain-Free Tempeh and Avocado

DAY 2

BREAKFAST Green Smoothie

SNACK Romaine Lettuce Boats Filled with Guacamole

LUNCH Romaine Salad with Avocado, Cilantro Pesto, and Grain-Free Tempeh

SNACK Romaine Lettuce Boats Filled with Guacamole

DINNER Lemony Brussels Sprouts, Kale, and Onions with Cabbage "Steak"

DAY 3

BREAKFAST Green Smoothie

SNACK Romaine Lettuce Boats Filled with Guacamole

LUNCH Hemp Tofu-Arugula-Avocado Seaweed Wrap with Cilantro Dipping Sauce

SNACK Romaine Lettuce Boats Filled with Guacamole

DINNER Roasted Broccoli with Cauliflower "Rice" and Sautéed Onions

DAY 4

BREAKFAST Green Smoothie

SNACK Romaine Lettuce Boats Filled with Guacamole

LUNCH	Longevity Leek Soup
SNACK	Romaine Lettuce Boats Filled with Guacamole
DINNER	Hemp Tofu-Arugula-Avocado Seaweed Wrap with Cilantro Dipping Sauce

DAY 5

BREAKFAST	Green Smoothie
SNACK	Romaine Lettuce Boats Filled with Guacamole
LUNCH	Creamy Cauliflower Parmesan Soup
SNACK	Romaine Lettuce Boats Filled with Guacamole
DINNER	Cauliflower "Fried Rice"

FREE DAYS

After you finish your five-day "fast," you can begin the free days portion of the plan and eat as much as you'd like of your gut buddies' favorite foods while you continue eliminating the foods that nurture bad bugs. If you go from the five-day "fast" back to your old habits, particularly sugar consumption, bad bugs can regrow quickly, undoing much of the progress you made.

On free days, you do not have to limit calories, but you should be mindful of your protein intake. As you read earlier, Dr. Valter Longo and I agree that you need only 0.37 gram of protein per kilogram of body weight. Therefore, you can easily meet your protein needs for an entire day with a scoop of whey protein powder, a couple of eggs, one protein bar, or about three ounces of pastured chicken or wild fish.

To make it simple, on your free days, I want you to focus on eating, at maximum, one roughly three-ounce serving of protein. You choose whether that means eggs at breakfast, a salad with tuna at lunch, or a small serving of wild fish or shellfish with dinner. Please start to view grass-fed and -finished beef as an occasional treat rather than as a diet mainstay. At your other meals, you'll get plenty of protein from veggies, nuts, mushrooms, and pressure-cooked lentils and by recycling the mucus in your gut. The latter sounds strange (and slightly gross), but it's true.

Other than that, feel free to enjoy any and all of the following foods.

Longevity-Promoting Acceptable Foods

Oils

Olive oil	Macadamia oil
Algae oil	MCT oil
Coconut oil	Avocado oil

Perilla oil
Walnut oil
Red palm oil
Rice bran oil
Sesame oil
Flavored cod-liver oil

Sweeteners

Stevia (SweetLeaf is my favorite)
Just Like Sugar (made from chicory root [inulin])
Inulin
Yacon
Monk fruit
Luo han guo (aka monk fruit; the Nutresse brand is good)
Erythritol (Swerve is my favorite as it also contains oligosaccharides)
Xylitol

Nuts and Seeds (½ cup per day)

Macadamia nuts
Pili nuts
Baruka nuts
Walnuts
Pistachios
Pecans
Coconut (not coconut water)
Coconut milk (unsweetened dairy substitute)
Coconut milk or cream (unsweetened, full-fat canned)
Hazelnuts
Chestnuts
Brazil nuts (in limited amounts)
Pine nuts
Flaxseeds
Hemp seeds
Hemp protein powder
Psyllium seeds or powder

Olives

All

Coconut Yogurt (plain)

Dark Chocolate

72% cacao or greater	(1 ounce per day)

Vinegars

All

Herbs and Seasonings

All except chili pepper flakes	Miso

Bars

Adapt Bar: coconut and chocolate	(www.adaptyourlife.com)

Flours

Coconut	Green banana
Almond	Sweet potato
Hazelnut	Tiger nut
Sesame (and seeds)	Grape seed
Chestnut	Arrowroot
Cassava	

Ice Cream

Coconut milk dairy-free frozen dessert (the So Delicious blue label, which contains only 1 gram of sugar per serving)

"Foodles" (my name for acceptable noodles)

Cappello's gluten-free fettuccine and other pastas
Pasta Slim shirataki noodles
Kelp noodles
Miracle Noodle brand pasta
Miracle Noodle Kanten Pasta
Miracle Rice
Korean sweet potato noodles
Palmini Hearts of Palm Linguine

Wine (6 ounces per day)

Red

Spirits (1 ounce per day)

Dark spirits like bourbon, scotch, dark tequila, dark rum, cognac, gin. Avoid vodka.

Fruits (limit all to their seasons except avocado)

Avocados
Blueberries
Raspberries
Blackberries

- Strawberries
- Cherries
- Crispy pears (Anjou, Bosc, Comice)
- Pomegranates
- Kiwis
- Apples
- Citrus fruits (no juices)
- Nectarines
- Peaches
- Plums
- Apricots
- Figs
- Dates

Vegetables

Cruciferous Vegetables

- Broccoli
- Brussels sprouts
- Cauliflower
- Bok choy
- Napa cabbage
- Chinese cabbage
- Swiss chard
- Arugula
- Watercress
- Collards
- Kohlrabi
- Kale
- Green and red cabbage
- Raw sauerkraut
- Kimchi

Other Vegetables

- Treviso, radicchio
- Chicory
- Curly endive
- Nopales cactus leaves
- Celery
- Onions
- Leeks
- Chives
- Scallions
- Carrots (raw)

Carrot greens
Artichokes
Beets (raw)
Radishes
Daikon radish
Jerusalem artichokes (sunchokes)
Hearts of palm
Cilantro
Parsley
Okra
Asparagus
Garlic
Mushrooms

Leafy Greens

Romaine
Red- and green-leaf lettuce
Mesclun (baby greens)
Spinach
Endive
Dandelion greens
Butter lettuce
Fennel
Escarole
Mustard greens
Mizuna
Parsley
Basil
Mint
Purslane
Perilla
Algae
Seaweed
Sea vegetables

Resistant Starches

Tortillas (Siete brand—only those made with cassava and coconut flour or almond flour)
Bread and bagels made by Barely Bread
Julian Bakery Paleo Wraps (made with coconut flour)
The Real Coconut Café Tortillas and Chips

In Moderation

Green plantains
Green bananas
Baobab fruit
Cassava (tapioca)
Sweet potatoes or yams
Blue or purple sweet potatoes
Rutabaga
Parsnips
Yucca
Celery root (celeriac)
Glucomannan (konjac root)
Persimmon
Jicama
Taro root
Turnips
Tiger nuts
Green mango
Millet
Sorghum "popcorn"
Green papaya

Plant-Based "Meats"

Quorn: Chik'n Tenders, Grounds, Chik'n Cutlets, Turk'y Roast, Bacon-Style Slices
Hemp tofu
Hilary's Root Veggie Burger (www.hilaryseatwell.com)
Tempeh (grain free only)

Pressure-Cooked Legumes (or Eden brand canned)

Lentils (preferred)
Black soybeans
Chickpeas
Adzuki beans
Other beans
Peas

Disease-Promoting, Life-Shortening Foods to Avoid

Refined, Starchy Foods

Pasta
Potatoes
Potato chips
Milk
Bread
Tortillas
Pastry
Wheat, rye, barley, rice, quinoa, soy, corn flours
Crackers
Cookies
Cereal
Sugar
Agave
Sweet One or Sunett (acesulfame-K)
Splenda (sucralose)
NutraSweet (aspartame)
Sweet'N Low (saccharin)
Diet drinks
Maltodextrin

Vegetables

Peas
Sugar snap peas
Legumes
Green beans
Chickpeas (including as hummus)
Soy products
Tofu
Edamame
Soy protein
Textured vegetable protein (TVP)
Pea protein
All beans, including sprouts
All lentils

Nuts and Seeds	
Pumpkin seeds	Peanuts
Sunflower seeds	Cashews
Chia seeds	

Fruits (some are incorrectly called vegetables)	
Cucumbers	Eggplant
Zucchini	Tomatoes
Pumpkins	Bell peppers
Squash (any kind)	Chili peppers
Melon (any kind)	Goji berries

Non–Southern European Cow Milk Products (these contain casein A1)	
Yogurt (including Greek yogurt)	Ice cream
	Frozen yogurt

Cheese	
Ricotta	Kefir
Cottage cheese	

Grains, Sprouted Grains, Pseudograins, and Grasses	
Wheat (pressure-cooking does not remove lectins from any form of wheat)	Farro
	Kamut
	Oats (cannot pressure-cook away the lectins)
Einkorn wheat	Quinoa

Rye (cannot pressure-cook away the lectins)
Bulgur
White rice
Brown rice
Wild rice
Barley
Buckwheat
Kashi
Spelt
Corn
Corn products
Corn syrup
Popcorn
Wheatgrass
Barley grass

Oils

Soy
Grape seed
Corn
Peanut
Cottonseed
Safflower
Sunflower
"Vegetable"
Canola

Acceptable Animal Protein Sources in Limited Amounts

Dairy Products (1 ounce cheese or 4 ounces yogurt per day)

Real Parmesan (Parmigiano-Reggiano)
French or Italian butter
Buffalo butter (available at Trader Joe's)
Ghee
Goat yogurt (plain)
Goat milk as creamer
Goat cheese, goat butter
Goat or sheep kefir

Sheep cheese and yogurt (plain)
Aged French, Italian, or Swiss cheese
Buffalo mozzarella
Casein A2 milk (as creamer only)
Organic heavy cream
Organic sour cream
Organic cream cheese

Fish (wild caught; 4 ounces per day maximum)

Whitefish, including cod, sea bass, redfish, red or pink snapper
Freshwater bass
Freshwater perch, pike
Alaskan halibut
Canned tuna
Alaskan salmon
Hawaiian fish, like mahi-mahi, opakapaka, ono

Shellfish (wild caught)

Shrimp
Crab
Lobster
Scallops
Calamari (squid)
Clams
Oysters
Mussels (farmed okay)
Abalone (farmed okay)
Sea urchin (uni)
Sardines
Anchovies
Smelt

Pastured Poultry (not free range; 4 ounces per day)

Chicken
Turkey
Goose
Duck
Pheasant
Quail

Pastured, non-soy or -corn fed, or omega-3 eggs (up to 4 daily), but limit whites, e.g., make an omelet with 4 yolks and 1 white)

Ostrich

Meat (grass fed and finished; 4 ounces per day maximum; once per week maximum)

Bison

Wild game

Venison

Boar

Elk

Pork (humanely raised or pastured)

Lamb

Beef

Prosciutto

Bresaola

Liver and other organ meats

Meal Plans

Here are a few daily menus based on the recipes later in the book to give you an idea of what to eat on your free days.

DAY 1

BREAKFAST	Blueberry Miso Muffins
LUNCH	Creamy Cauliflower Parmesan Soup with a side of Bitter Green Salad with Walnut Blue Cheese Dressing
DINNER	Roasted Broccoli with Miso Walnut Sauce, Mushroom and Thyme Braised Tempeh over cauliflower rice
SNACKS & DESSERT	½ avocado with Miso Sesame Dressing; piece of in-season fruit

DAY 2

BREAKFAST	Toasted Millet "Grits" with Spicy Eggs
LUNCH	Roasted Broccoli with Miso Walnut Sauce, Mushroom and Thyme Braised Tempeh over cauliflower rice
DINNER	Lentil Broccoli Curry with Ginger Coconut Cauliflower "Rice"

SNACKS & DESSERT	Basil Lentil "Pâté"; Mexican Chocolate "Rice" Pudding

DAY 3

BREAKFAST	Blueberry Miso Muffins (leftover from day 1)
LUNCH	Lentil Broccoli Curry with Ginger Coconut Cauliflower "Rice"
DINNER	Spinach Salad with Lentil-Cauliflower Fritters
SNACKS & DESSERT	½ avocado with Miso Sesame Dressing; Mexican Chocolate "Rice" Pudding

BRAIN-WASH DAYS

As you read earlier, your body needs to finish digesting your last meal at least four hours before you go to sleep in order for your glymphatic system to thoroughly wash out your brain at night. This is how you can avoid the toxic buildup of amyloid in the brain that can lead to degenerative disease. To get a thorough brain wash, I recommend that you skip dinner once a week if you are in good health, more often if you are already suffering from a degenerative disease. On a brain-wash day, you will simply eat as if it were any other free day and then stop eating after lunch. Simple. You can even condense all three meals into the first half of the day if you wish, eating your last meal no later than 4:00 p.m.

This will also allow you to take advantage of the benefits of intermittent fasting. In a new study on mice at the National Institutes of Health, the mice that went the longest between meals had the greatest longevity outcomes no matter what they actually ate.[5] So for the best results, don't just skip dinner but also make sure that eighteen hours pass before your next meal. In fact, limiting your eating to a six-hour window daily stimulates autophagy, your cells' recycling program.[6] When autophagy occurs at the intestinal lining, stronger, healthier cells reinforce the gut barrier.[7] So by just skipping dinner and delaying your breakfast (break-fast, get it?) you get a cleaned-out brain, younger cells, and a renewed gut barrier. That's certainly worth missing one meal a week!

OPTIONAL CALORIE RESTRICTION DAYS

If you want to go all in on calorie restriction, I recommend my version of the 5:2 diet. On this plan, you break your week up into five free days and two calorie restriction days when you eat 600 calories per day. If you do the math (don't bother, I did it for you), this equates to cutting 540 calories every day for a week, giving you the benefits of full-time calorie restriction and weight loss of a pound a week, a healthy and safe rate at which to lose weight.

Though your gut buddies don't care which days you choose for your calorie restriction days, I strongly advise doing it on Monday and Thursday. Why? On Monday you are coming off the weekend, when it's possible that you overindulged, so it's a great time to cut back. Then you have two free days before you arrive at Thursday, which is just before the weekend—another great time to cut back. Of course, you can change these days each week or however often you like to accommodate your schedule.

So what does 600 calories look like? There are approximately 600 calories in three Quest bars or seven to eight hard-boiled eggs. In fact, the "egg diet" really works, as a recent human trial showed! I'm not kidding—several of my patients successfully eat this way. Salads work really well on these days, but remember that one tablespoon of olive oil will cost you 120 calories. Or how about the Drinking Man's Diet? Seriously, that popular diet from the 1960s and '70s by Robert Cameron, whom I mentioned earlier, suggested drinking three-quarters of a bottle of red wine per day, a little bit more than our 102-year-old friend Luigi Cornaro drank daily. And Cameron made it to age 98! Imagine what could have happened if he had cut back a trifle? Surprisingly, there are only about 525 calories in a full bottle of wine, so you could sip on that bottle throughout the day and pair it with one hard-boiled egg for good measure! (Just joking, but you get my point.)

More practically, I recommend eating lots of raw and cooked veggies with a little bit of olive oil and a small amount of a concentrated vegetable or nut protein on these days to get maximum nutrition while benefiting from calorie restriction.

OPTIONAL INTENSIVE CARE CLEANSE

As you read earlier, cancer cells and certain immune cells share a little-known vulnerability: they cannot efficiently utilize fat as a fuel. Instead, they must go through a highly inefficient process known as sugar fermentation to derive energy. If you have been diagnosed with cancer or are suffering from an acute autoimmune disease, Parkinson's disease, or dementia, the intensive care cleanse provides your mitochondria with the fuel they need while starving the cells that are responsible for your condition. You can use this as a kick start to the rest of the program as a three-day cleanse or stay on it for life if you find that it helps your gut buddies and their sisters thrive.

For this program, you'll follow the list of your gut

buddies' favorite foods on pages 311–317 that you will eat on free days with the following changes.

Eliminate all fruits and seeded vegetables—their fructose is too tempting for cancer cells. Consume absolutely no fruit except for avocados, green bananas and plantains, green mangoes, and green papayas.

Opt for medium-chain fatty acids such as MCT oil, coconut oil, or red palm oil or for the short-chain fatty acids in butter or ghee as oil/fat sources, but continue to use olive oil as your main source of fat. Think of other foods as fat delivery devices. Eat as much of these fats as you can.

Eat macadamias as your preferred nuts, with smaller amounts of other nuts.

You can still treat yourself to extra-dark chocolate, but be sure it contains at least 90% cacao. Lindt makes a good 90% cacao chocolate bar that is widely available, and as of this writing, Trader Joe's carries a 100% cacao bar with cacao nibs called Montezuma's (and no, there's no revenge).

Eat no more than 2 ounces of animal proteins—the size of a quarter of a deck of cards—a day, preferably in the form of wild fish, shellfish, and mollusks. If you have cancer, avoid animal protein completely, as certain amino acids in animal protein fuel cancer cells.

Egg yolks are virtually pure fat and one of two fats your brain needs to function properly. Try a four-yolk, one-whole-egg omelet cooked in coconut oil or ghee and filled with sliced avocado, mushrooms, and onions and sprinkled with turmeric and black pepper. Douse it with more ghee or macadamia oil, perilla oil, or olive oil before serving.

Vegans can have half a Hass avocado with a douse of coconut or olive oil. Hemp seeds are a good source of fat and plant protein and can be used in shakes or sprinkled on salads or veggies. Walnuts have the highest plant protein content of the nut choices.[8]

Putting it all together, let's take a look at what one month on the Longevity Paradox program looks like.

WEEK 1

5 fast-mimicking days followed by 2 free days

WEEKS 2, 3, AND 4

4 free days, 2 calorie restriction days of your choice, and 1 brain-wash day; or

6 free days and 1 brain-wash day; or

5 free days and 2 brain-wash days; or

add as many days as you wish of the optional intensive care cleanse

It's up to you to customize the program to suit your needs. If you are young and in good health, feel free to start slowly. This is a program that you can live with, literally and figuratively. However, if you are already suffering from dementia, type 2 diabetes, an autoimmune condition, or any other disease of aging that you now know is actually the result of bad bugs ruling your inner kingdom, you may need to follow the protocol to the letter.

Your life is literally in your hands—or, shall I say, in your mouth and gut. But although your diet is the number one factor that will determine which type of gut bugs are able to thrive in your body—and therefore how long you'll thrive—it's not the only one. Your daily habits, such as the way you exercise, the temperature of your shower, and even those who you spend time with, all play a role in determining your life and health span. And that leads us to the next phase of the Longevity Paradox program.

Chapter 10
The Longevity Paradox Lifestyle Plan

As you know, one reason calorie restriction is beneficial is that it temporarily stresses your cells, and a little bit of stress is a good thing: it sends a signal to your cells that they should prepare for an impending threat to your and their survival. This forces them to toughen up and kills off any cells that can't be strengthened and therefore aren't likely to survive the onslaught. It is one of the most beneficial things you can do to promote your health and longevity.

That's why the next pillar of the Longevity Paradox program involves stressing your cells not only through your diet but also through your lifestyle choices. As you begin this journey, it's important to keep in mind that the more you stress your body, the more time you need to recover, or you risk stressing yourself *too* much and causing more harm than good. So getting plenty of sleep and taking time to relax or meditate are also essential components of this program. Alternating between periods of stress and rejuvenation is another cycle that will benefit your gut buddies and help you achieve a long health span.

With this in mind, I've broken down the lifestyle program into two parts: first, the habits that will stress and strengthen your cells, and second, the habits that will allow them to recover. Together, these simple lifestyle modifications will leave you—and your gut buddies—feeling better than ever.

PART 1: CONQUER STRESS

Exercise is one of the most commonly practiced forms of hormesis. Every time you work out, you create tiny tears in your muscles. When your muscles repair themselves, they become stronger and bigger. And as you read earlier, your gut buddies also benefit when you exercise and repay you by fixing up their home. They especially like it when you exercise against gravity because this stresses—and therefore strengthens—more of your muscles.[1]

If you are concerned that you're not in great shape and can't safely start a weight training protocol, have no fear. Jack LaLanne, "the Godfather of Modern Fitness," taught me that you need to do only two simple exercises to develop and maintain strength. Those two exercises are squats (or any type of deep knee bends) and planks or push-ups. Both exercises work against gravity, and together they stress every major muscle group in the body. Anyone at any fitness level can do them, and just a small investment of time will yield meaningful results.

My five-minute exercise plan incorporates these movements and three others you'll recognize for a complete, well-rounded workout that provides the amount of stress your muscles need to stay strong and prevent wasting, or loss of muscle mass with age. You have no excuse to avoid doing this. Start off by completing this circuit twice a day or whenever you feel the need to get up and get moving, especially if you spend a large portion of your day sitting. It will give you an instant burst of energy while strengthening all the muscles—and cells—in your body.

Step 1: The Longevity Paradox Exercise Plan

First Minute: Jog in Place

I like to call this the Standing Trot. This is not really a full jog; it's a trot. Don't overdo it. Remember, *if you're getting hot, it's not a trot.* You want to wake yourself up, not wear yourself out! Just do a nice, easy trot for one minute. If this is too much for you, an easy modification is to do it seated. Move your legs and arms as though you're running while sitting upright in a chair. If you feel silly while doing this, just keep going anyway. Talk to strangers walking by your table at Starbucks. Have a chuckle about it if you need to. Heck, I laugh at myself all the time (hey, watch those comments!), and I believe it keeps me young. In fact, I've become friendly with "the Laughing Yogi," who, you guessed it, laughs out loud during yoga and has taught me to do so as well.

Second Minute: Classic Crunches

If your abs get stronger, you'll stay young longer. A strong core is essential to maintaining mobility and eliminating back pain. To do a crunch with proper form, lie on your back with your spine straight and your knees bent, arms pointed toward your feet, and focus on lifting your head and shoulders up using your abs, not your neck or your arms. Pull in your lower abdominals, around your belly button, and slowly peel your spine off the floor vertebra by vertebra; speed is not your friend here. You don't need to sit all the way up. Just make sure you feel the engagement of your abdominal muscles as you lift. Repeat as many times as you can, keeping good form, for one minute. If this bothers your back, try a supported crunch instead. Just rest your calves on a chair or bench with your knees bent at a 90-degree angle. Do your crunches in this position to ease the pressure on your spine. Still too much strain on your neck? Do the crunches the old-fashioned way: support your head with your hands, but don't pull up with them.

Third Minute: Plank

This is one of Jack LaLanne's favorites—and mine. Planks are great because they engage all of your muscles at once, yet they don't require any movement! To do a plank, simply get into the top of a "push-up" position and hold it for one minute. Keep your back straight, your butt slightly high, your abs pulled in, and your hands directly below your shoulders with your arms straight. If this is easy for you, simply do push-ups for that minute instead. But if it's too difficult for you at first, don't worry; it is perfectly normal not to be able to hold a plank for a full minute without stopping. Take a break whenever you need to and then get back into position. Or if you need to take some weight off, lower your knees to the ground. Still too hard? Perform the plank resting on your elbows with your forearms out front. Just make sure to keep your upper body and core engaged.

Fourth Minute: Squats

Now we move on to another of Jack LaLanne's favorites. To this day, I do squats every morning and every night while I'm brushing my teeth! I have nothing better to do during this time, and this way I keep my gums healthy and strengthen my lower body and core at the same time.

To do squats properly, stand with your feet parallel to each other a little wider than hip width apart. Inhale, draw in your abdominal muscles, and slowly bend your knees while keeping your chest forward and your head lifted. Bend as deeply as your mobility will allow, then return to a standing position by engaging your gluteal muscles. Do as many repetitions as you can in one minute, focusing on keeping your legs parallel and your abs engaged the whole time. Feeling unbalanced? Just hold on to a counter or the back of a chair with one hand.

Fifth Minute: Meditation

You thought this was going to be another type of movement, didn't you? Nope!

You're done moving, and it's time to bring your heart rate back down, relax, and sweep away the mental clutter. Don't be tempted to skip this step and move on with your day. Remember, your gut buddies love to meditate, so treat them to one minute (or more if you have time!) of complete relaxation.

Start by either sitting up straight or lying on your back, and focus on inhaling deeply through your nose and exhaling completely through your mouth. With each exhalation, think about relaxing all of the muscles in your body. Try relaxing your feet first, then your knees and thighs, then your back, then your arms and hands, then your neck—you get the idea. After just a minute, you'll be refreshed, alert, and energized—and so will your gut buddies. And no, you don't need to let your mind go blank; that just adds to many people's tensions!

A Prescription to Play!

Going outside and moving your body is so important that I often literally write my patients a prescription to get a dog. Many of them have thanked me later, saying it was the best thing they've ever done. Dogs not only force you to get outside and get walking, but they get dirty. Their paws get muddy, and their coats pick up pollen spores and leaves. Many people see this as a negative at first, until I explain that exposure to the bacteria in dirt is good for their gut buddies. In fact, research shows that people who live with dogs have far more diverse microbiomes than do non–pet owners.[2] And kids who have dogs have far fewer allergies than those who don't. Talk about man's best friend!

Whether you have a dog or not, get outside as much as you can and get moving. I like to take a ten-minute walk after each meal. You don't have to strain yourself. Simply move. Walk. Go outside and weed the garden or mow the lawn. Your body is designed to be active, not to sit for long periods of time. If possible, take your daily walks up and down hills (against gravity) to stress more muscles and reap even greater benefits.

If and when you're ready for more, I highly recommend another simple exercise that anyone can do at any age—and I guarantee you'll have a gigantic smile from ear to ear while doing it. It's called rebounding, but it is really just bouncing on a minitrampoline. I got a rebounder at Target for less than thirty dollars, and I love bouncing on it. It's easier on the joints than jogging—or even walking—on a hard surface. Not to mention that it's great for your lymphatic system, which carries vital nutrients to all of your cells. My good friend and patient Tony Robbins bounces on his minirebounder just before he bounds onto the stage. It's located right next to the stairs. No wonder he looks (and is) so energized!

To get started, stand on your minitrampoline with your feet shoulder width apart. Bend your knees and lightly bounce up and down. Do this for one minute. Rest for one minute. Repeat two more times. Easy! I do this once a week, and it makes me feel like a kid. Feeling wobbly? Many rebounders come with a pole or handle for balance purposes. So take this as your prescription to play—it will actually make you young again.

HIIT for Fun

Another exercise option that provides big benefits for a small investment of time is high-intensity interval training (HIIT). HIIT burns more fat for longer than traditional exercise does. Plus, whereas traditional aerobic exercise can weaken the immune system, HIIT doesn't have that effect. As little as ten minutes of HIIT three times a week can confer a variety of health benefits, especially for those who were previously inactive. So if you've been waiting for an excuse to start a new exercise routine, here it is. HIIT is also a highly enjoyable form of exercise. In fact, research shows that it elicits greater feelings of happiness than traditional forms of moderate-intensity exercise do because it leads to greater neurotransmitter release.[3] And of course, as you know well by now, your gut buddies are the ones producing the precursors to serotonin and other "happy-making" hormones, so it's safe to say that they enjoy HIIT, too!

On the Longevity Paradox program, I recommend doing a ten-minute HIIT routine three times a week in addition to your five-minute daily circuit. You can choose any form of exercise you prefer: walking, running, cycling, spinning, or just doing jumping jacks. For thirty seconds or so, work as hard as you can, followed by roughly an equal amount of recovery time. As you get stronger, you might extend your bursts of intensity to a minute. Just be sure to give yourself adequate recovery afterward. Continue until your ten minutes are up, and you'll be happier and healthier—and so will your gut buddies.

Step 2: Cook Your Cells—Just a Little

As you read earlier, I first studied heat-shock proteins when performing heart surgery. I needed to keep the heart from being injured during beating-heart surgery by temporarily cutting off blood flow to certain regions of the heart in order to perform the surgery, so I would clamp down a blood vessel for two or three minutes at a time and then release it. My colleagues and I found that when we did this, the heart cells would produce heat-shock proteins to protect the heart against the stress of having no oxygen supply. Once the heat-shock proteins developed, we could reclamp the blood vessel—this time for ten minutes at a time without having to worry about damaging the heart, which could now withstand longer periods without blood flow because it was so well protected.

This is a perfect example of hormesis in action. Your cells get the message that tough times are coming (in this case because of a lack of blood flow) and that they had better toughen up. So they develop heat-shock proteins as a means of protection. These proteins tell any cells that are not carrying their weight to self-destruct, leaving only healthy, fresh cells once the tide has turned and all is well again.

But heat-shock proteins protect you against all kinds of threats, not just a lack of blood flow. As their name implies, your cells produce these protective proteins when temporarily stressed by extreme temperatures. So once a week or so, try to spend some time in a sauna or a steam room, expose yourself to a near-infrared or red-light sauna, take a hot yoga class, or come visit me in Palm Springs in the summer![4] If these options aren't realistic, simply take a nice hot bath. Recent studies have shown that a hot bath relieves mild depression better than antidepressants. To avoid overstressing your cells, get into the bathtub when it's fairly warm, and then keep letting out some of the water and adding more hot water. You'll achieve the same hormetic effects as long as you're sweating.

Step 3: Toughen Up for the Winter

Exposure to cold temperatures has a similar effect as heat exposure, signaling to your cells to take up a defensive posture to survive a long, brutal winter. Many hibernating animals survive near-freezing temperatures during the winter by producing a protective compound when they are exposed to cold temperatures. Is this why rodents that hibernate live on average twice as long as those that don't? Or is it because hibernating rodents have a far lower metabolic rate, which, as you read earlier, promotes longevity? I suspect it's a little bit of both. Exposure to cold temperatures also stimulates your gut buddies to produce more of two beneficial neurotransmitters, GABA and serotonin, both of which help extend life span.[5]

To take advantage of these benefits, I recommend taking a daily "Scottish shower." This may not sound like fun, but it will certainly wake you up in the morning and force your body to stay energized and activated throughout the day. To do so, start your shower with warm water as usual and then gradually cool the water down. By the last couple of minutes of your shower, you should be running nothing but cold water. I promise you'll get used to this fairly quickly. And just think about how much money you'll save on your hot water bill!

If you're still not convinced, think about the fact that your skin buddies also love a Scottish shower. Hot water strips out the nourishing oils they produce for your skin and hair. Cold water, on the other hand, leaves these important oils intact. This gives you smooth, bright skin and full, shiny hair in addition to a boost in energy and a longer, healthier life. Still too extreme? Buy a cold vest—literally, a vest with removable cold packs—and wear it for a few hours a day. There are even ones that you can wear under a suit or dress.

PART 2: REJUVENATE

Okay, are you done being stressed? Good, then it's time to move on to the rejuvenating aspects of the Longevity Paradox program. After all, if you overdo it on the stress and don't take time to recover, not only will you fail to benefit from all that stress, but you'll actually cause your body harm by not allowing the stressed cells to return to normal and repair the microdamage that has been done—the exact opposite of what you want to achieve. So don't skip these important steps even if they feel unnecessary. They are an important part of your long-term health and happiness.

Step 1: Prioritize Sleep

It's no secret that sleep is important for a long, healthy life, as my friend Arianna Huffington so eloquently showed in her book *The Sleep Revolution.* Deep sleep is when your glymphatic system "washes" your brain, scrubbing it of junk and debris so it doesn't build up an accumulation of the amyloid plaques that can eventually lead to Alzheimer's and other neurodegenerative diseases. You're already skipping dinner at least once a week in order to make sure you have enough blood flowing to your brain as soon as you get to sleep in order to complete that important cleanse.

But once a week is just a start. Your brain needs adequate sleep every night, not just on your brain-wash evenings. It turns out that losing just one night of sleep results in amyloid increases of about 5 percent in the brain's thalamus and hippocampus, the regions of the brain that are especially vulnerable to damage in the early stages of Alzheimer's disease.[6] Ouch! This suggests that chronic sleep deprivation may be a leading cause of Alzheimer's.

If that's not enough to get you to bed on time, how about this: lack of sleep will also cause you to gain weight as you age. The two hormones that control hunger and satiety, ghrelin and leptin, respectively, are very sensitive to sleep duration. When college students were put into a sleep lab and allowed to sleep for eight hours, the following morning they had high levels of leptin (which tells you that you're full) and low levels of ghrelin (which tells you to keep eating). The next night they were awakened after just six hours. This time their ghrelin levels were high and their leptin levels were low, making them hungrier and more difficult to feel sated.[7] I can assure you from personal experience as a surgical resident with thirty-six hours on and twelve hours off for months at a time that getting very little sleep makes you gain a lot of weight!

This, of course, corresponds to the yearly cycle. The

long days of summer, with their shorter nights, stimulate the body to eat and to start storing fat for the upcoming winter. To put it simply, when you get a lot of sleep, your body thinks it's winter and that it should burn up all the fat that is stored on your body. But when you miss sleep, your body thinks it is summer and that it should start storing fat to prepare for winter.

This is another reason that hibernating animals live longer than ones that don't hibernate. They are living off of that stored fat as ketones, which burn cleanly. Think of ketones as clean-burning natural gas, while proteins and sugars are dirty diesel fuel. With the former there's no need for a catalytic converter! You want your body to think it's winter year-round so that it will keep you strong, lean, and alive.

Sunlight is, of course, one of the sources that signals to your body the time of day and time of year. When your body is exposed to sunlight, key receptors in your retinas are activated, which stimulates your gut buddies to produce hormonal and chemical messengers that make you feel either sleepy or alert. So, for example, if it is dark out, your retinas pick up on the message that it is nighttime and send an alert for your body to create the neurotransmitter melatonin, which helps you sleep. This is commonly referred to as one of your circadian rhythms.

Unfortunately, the blue spectrum in light stimulates the alertness and feeding cycle that your body associates with summer. With more hours of daylight during the summer months, it is time to stay up later and eat more food to store for the coming winter. It's not only the intensity of blue light but also its duration that determines your seasonal patterns of sleeping, waking, and eating.

We humans have special receptors in our eyes that process blue-spectrum light, which tells our bodies to wake up and be alert. Before screens and fluorescent lights took over the world, the sun was our only source of blue light. So for most of human existence, we saw blue-spectrum light only when the sun was out during the daytime. The human body evolved to use this daytime blue-light source to set its circadian rhythm. But now, screens and lightbulbs emit huge amounts of blue light at any time of the day or night. This interferes with your gut buddies and your circadian rhythm, which can in turn lead you to age more rapidly.

In order to get the sleep your body needs, it's important to reestablish your circadian rhythm and get back in sync with the natural ebb and flow of daylight. Here are some ways to do this.

- Reduce your exposure to blue-spectrum light at night. Once the sun goes down, turn off all screens. It may be unrealistic to do this all the time, but try your best at least a few nights a week. Check out apps such as f.lux that reduce your computer screen's amount of blue light the rest of the time.

- Get a pair of blue light–blocking glasses and wear them after the sun normally goes down while you are at home reading, watching TV, or on your computer that you were supposed to shut down. (I'm watching you!) These glasses can protect your eyes from overexposure to blue light. In fact, did you know that blue light–blocking glasses were originally developed for NASA astronauts to use, since the sun's rays are even stronger in space? If they work for astronauts, surely they can work for you.
- Try taking time-released melatonin supplements to reset your body clock if it's been thrown off. The pills usually come in 3- to 5-milligram doses—and that's plenty. Time-released melatonin is particularly useful when it comes to adjusting from jet lag and resetting your sleep-wake cycle when your work schedule is inconsistent.
- Keep a consistent sleep/wake rhythm. Go to bed at the same time each night, or at least as often

as possible. Make sure you will sleep a full eight hours before you need to wake up. Remember, this is not a luxury; your brain, your body, and your gut buddies all depend on it.

- Last, try to be consistent in your sleep schedule. Sleeping in on the weekend often doesn't make up for time lost during the week. Want some help? I know I sound like a broken record, but get a dog! Believe me, it won't let you sleep in and it usually gives you a signal when it's bedtime.

Now that you've recovered from all of that self-imposed stress, it's time to move on to the next part of the Longevity Paradox program, which I'd be willing to bet isn't what you expect.

Step 2: Kiss and Connect

There is one thing the people in the long-lived Blue Zones have in common that we have not yet addressed, and that is that they all live in extremely tightly knit communities. In fact, research shows that most centenarians, regardless of their culture or country of origin, have a strong social and spiritual support system.

This reminds me again of Edith ("Michelle"), the age-defying beauty you read about earlier, who recently passed on while she was very young, right before turning 106. When she was a youthful 101-year-old, she fell in the bathroom and broke her hip. I honestly thought that was going to be the end. There was a period of about six months when her decline was obvious. For the first time, she became forgetful and began showing signs of her age. But I underestimated her.

Edith had a robust social network. She was always out and about with her adorable little Pomeranian. That dog got her back out onto her feet after the fall, and soon she was making plans to meet friends for lunch. Soon her social calendar was full of appointments, reasons to get out of bed, connections to be made and nurtured. She enjoyed another five good years after that fall, her mental status recovered fully, and I believe that her strong network of social and emotional support had a lot to do with that.

The same principle is true for the long-lived Seventh-Day Adventists of Loma Linda. They live within a tight-knit community that provides both practical and spiritual support to all its members. We see the importance of community even in the animal kingdom: naked mole rats work together to dig tunnels and find tubers to feed to their queen. They have the same fascinating social structure as bees, one that all other long-lived societies share, whether it's based on a family structure, a religion, or a village.

Perhaps this is why so many of my male (and increasingly female) patients immediately see their health begin to decline as soon as they retire. They lack their daily workplace social structure and become isolated, which ages them. In our culture, older folks tend to become isolated and depressed, but social connections are essential for good health. As a result, we are seeing an epidemic of loneliness that coincides with our decreasing health spans and life spans. It might be tempting to dismiss this as coincidence or purely a mental phenomenon, but it is actually, as you might suspect, tied very closely to the health of your gut buddies.

Think about it: the more people you come into contact with, the more bacteria you share—just like the dog that licks your face and spreads its bugs all over you (I hope your colleagues and book club members don't greet you that way, but hey, maybe they should! As I travel in France, Italy, and the rest of the Mediterranean, home to the longest-living people on Earth, men and women kiss each other's cheeks upon greeting). This idea also ties back to the study you read about earlier, which showed that people who live together often have the same types of health problems—not because they share genes but because they share microbiomes—and it explains why people tend to become obese if their close contacts are obese. It's not merely because people choose friends who share similar lifestyle habits—it's because they share microbiomes with their friends, and their gut bugs play a huge role in influencing their weight.[8,9]

Your chance of becoming obese increases by 57 percent if you have a friend who becomes obese in a given interval. Among pairs of adult siblings, if one sibling becomes obese, the chance that the other will become obese increases by 40 percent. If one spouse becomes obese, the likelihood that the other spouse will become obese increases by 37 percent. Notably, these effects are not seen among immediate neighbors. And the siblings do not become obese at the same time because they share the same genes; it's because they share the same gut bugs.[10]

It needs to be said: be careful with whom you share bacteria! Thankfully, our gut buddies help us even with this, at least when it comes to selecting our mates. There is now evidence to suggest that kissing, which is universal among humans and other great apes, is more than a chance to exchange bodily fluids pleasantly. In another case of "fact is stranger than fiction," when we kiss another person, we are really sampling his or her bacterial mixture to see if it is compatible with our own.[11] If it is, our microbiome will tell us that it's a match by stimulating the production of feel-good hormones that make us want to kiss him or her again. (I hear the theme from *The Twilight Zone* playing whenever I think about this. I also think of all the pickup lines you single readers now have at your disposal regarding information sharing and microbiome research back at your apartment.)

So, what does this mean for you, especially if you're already married? Get out there and spread those germs around. Seriously! Join a group—any group—whether it's a book club, a workout group, or even Parents Who Refuse to Drink Alone on a Friday Night. Volunteer at a local organization that needs your help. Return to your spiritual roots or explore new roots. And for goodness' sake, get a dog, already!

I'm tempted to say that humans are social creatures, but by now it's clear that it is more accurate to say that your gut buddies are the ones that really want to be social and be exposed to new friends and acquaintances. Do them this favor while following the rest of the Longevity Paradox lifestyle plan, and they'll repay you in spades. I guarantee it. I've seen in scores of patients that "super seniors" are motivated to be with others, act as a source of knowledge for younger generations, and literally become the pillar of their families or communities. After all, what good is a long, healthy life if you can't spend it with the people you love? Human connection really seems to drive successful aging—not to mention the motivation to stay alive and vibrant and make sure your dog gets its walk twice a day!

Chapter 11
Longevity Paradox Supplement Recommendations

THE MEANING OF "SUPPLEMENT"

Many people still believe that there is a supplement magic bullet—in other words, that one or more supplements will somehow correct their ongoing reliance on the typical Western diet, as well as cause all their health issues to magically reverse course and heal their bodies. Moreover, if you scan the Internet, you will find wild claims that taking one particular supplement will make you immortal or almost. I can assure you that this is nonsense, and I say that because I have witnessed this misconception in my patients' blood work far too many times over the last eighteen years. However, if you embark on the Longevity Paradox program, many of the supplements that follow can and will provide measurable benefits. I have presented studies on such benefits at national and international medical conferences. Remember, true to their name, supplements enhance the results of the Longevity Paradox program—but they are not shortcuts.

Several of my colleagues in longevity take the prescription drug metformin (page 70) and some at least talk about and may secretly take the organ transplant antirejection drug rapamycin (Sirolimus), the latter for its known direct suppression of mTOR. I do neither, preferring the natural methods I have elucidated in this book and using other natural supplements to mimic these drugs' actions. A disclaimer: I own and operate my own supplement company, GundryMD, but in no way am I suggesting that you need to buy my products. I have combined many of my favorite nutrients together in formulas for GundryMD.com. but I also share the names of other brands I like, as well as the supplement dosage, so you can find whatever works best for you and your budget, either online or in your local health food store.

I used to tell my patients that supplements made expensive urine. That was before I started measuring the effects of vitamins, minerals, and plant compounds such as polyphenols, flavonoids, and other phytonutrients on my patients' inflammation biomarkers. I can now reliably discern when patients have changed their supplement regimen or even changed brands, based upon these tests.[1]

Our hunter-gatherer ancestors consumed more than 250 different plants annually on a rotating, seasonal basis. Those plants' roots delved deep into six feet of organic loam soil, which teemed with bacteria and fungi to create an amazing combination of minerals and phytochemicals within the plants' tubers, leaves, flowers, and fruits. The meat and fat from the animals that our forebears killed and ate also contained those phytochemicals, because the animals they ate also ate those plants.

Let's say that you eat organic food, you eat seasonally, you frequent your local farmers' market, you consume wild-caught seafood, you limit your consumption of pastured chicken and eggs, you limit your consumption of grass-fed meats, and you sprinkle aged cheeses from A2 cows, as well as from sheep and goats, on your food. You eat pressure-cooked lentils; you throw mushrooms on everything. You skip meals, you have a brain-wash night weekly. Isn't that enough? As the lab tests on many of my patients who are faithful organic eaters show, getting all of the nutrients you need simply cannot be done in our society without taking supplements. Unfortunately (or perhaps fortunately), you do not live in Okinawa in the 1940s, on Kitava, or on a remote island off the coast of Greece; you get my drift.

So here are a few of the supplements that I recommend considering. The first two—vitamin D3 and the B vitamins—are essential for everyone.

Vitamin D3

Most Americans have very low levels of vitamin D3. About 80 percent of the patients in my practice were vitamin D deficient when they first enrolled, including 100 percent of my autoimmune and lectin-intolerant patients. I have been shocked by how much supplementation some of my autoimmune patients need to get their vitamin D blood levels up to what I consider normal, which is 70 to 120 ng/ml for serum 25-hydroxyvitamin D, the active form of vitamin D in your body.

Because I measure vitamin D levels every three months, I can be aggressive with replacement, but if you are just beginning this program, please take just 5,000 IU of vitamin D3 daily. For autoimmune disease, start with 10,000 IU a day. In the last eighteen years, I have yet to see a case of vitamin D toxicity. In fact, I doubt that it exists.

The B Vitamins, Especially Methylfolate and Methylcobalamin

Many of the B vitamins are produced by gut bacteria, so if your gut rain forest has been decimated, it is likely that you are deficient in both methylfolate (the active form of folic acid) and methylcobalamin (the active form of vitamin B12, sometimes called methyl B12). Moreover, more than half of the world's population has one or more mutations of the methylenetetrahydrofolate reductase (MTHFR) genes (it is possible to have single or double mutations of the two most common genes), which limits their ability to make the active forms of both vitamins. Many people, including myself, call the MTHFR mutation the Mother F'er gene, thanks to how the acronym looks, but that's not how I refer to it in TV interviews! The good news is that by swallowing a 1,000-microgram methylfolate tablet each day and putting 1,000 to 5,000 micrograms of methyl B12 under your tongue, you can combat the genetic mutation. Since you have about a 50 percent chance of carrying one or more of these single or double mutations, I think it is worth taking the active forms of methyfolate and methyl B12 just in case. Although they will not hurt you, if you are one of the few with one or both of the double mutations, you may notice increased excitability or conversely depression.

Why should you take these B-vitamin supplements? Simply put, they contribute a methyl group to an amino acid called homocysteine in your bloodstream and convert it to a harmless substance. An elevated homocysteine level is correlated with damage to the inner lining of your blood vessels that is on a par with the damage caused by elevated cholesterol levels. Methyl groups (–CH3) are responsible for turning genetic switches on or off as well. These B-vitamin supplements almost always lower homocysteine levels to within normal range.

The G7

Years ago, when *Dr. Gundry's Diet Evolution* was first published, I was asked to name the seven most important classes of supplements that I felt everyone should have in his or her armamentarium to help achieve great health. We termed this the G7, in reference to the meeting of heads of state to determine the future course of the world (and in reference to the first letter of my last name). We shortened it to G6 for *The Plant Paradox*, but I am now re-adding a seventh for longevity. Here is my new G7 list.

Polyphenols

Perhaps the most important class of compounds missing from your diet is the plant phytochemicals called polyphenols. Plants design these compounds and concentrate them in their fruit and leaves to resist insects and protect against sunburn (yes, fruit gets sunburned), so polyphenols provide you with a host of beneficial effects when metabolized by your gut bacteria. Fun fact: those beautiful fall colors in leaves are the polyphenols that were always there, hidden by the dark green chlorophyll. Another fun fact: the leaves of a plant generally have more polyphenols than the fruit does. For instance, olive and apple tree leaves have more polyphenols than their olive or apple fruits, one of the reasons that olive leaf extract provides more of the benefits associated with olive oil. These benefits include blocking the formation of the atherosclerosis-causing trimethylamine N-oxide (TMAO) from the animal proteins carnitine and choline and actively dilating your blood vessels. These compounds are so important that I formulated my own blend called Vital Reds. The product combines thirty-four different polyphenols, as well as my favorite probiotic, a spore form of *Bacillus coagulans* (BC^{30}), into a powder that mixes easily with water.

However, as all my patients know, I don't even sell my own products in my offices, choosing instead to point out alternative sources of polyphenols. Some of my favorite polyphenols in supplement form are grape seed extract, pine tree bark extract (sometimes marketed as pycnogenol), and resveratrol, the polyphenol in red wine. You can find supplements at Costco, Trader Joe's, and Whole Foods and online. My suggested doses are 100 milligrams of both grape seed extract and resveratrol and 25 to 100 milligrams of pine tree bark extract a day. Other great additions are green tea extract, berberine, cocoa powder, cinnamon, mulberry, and pomegranate, many of which (and many more) are in Vital Reds but can also be taken separately.

In my opinion, the best resveratrol product available is Longevinex, which I have taken for at least eleven years. I have no relationship with the company but am impressed with its research. Resveratrol and other small molecules such as quercetin, which are present in red wine, are responsible for the activation of the SIRT1 gene, which in turn suppresses mTOR. You can find it at www.longevinex.com.

Green Plant Phytochemicals

Without a doubt, you cannot eat enough greens to satisfy your gut buddies, a fact that you will soon witness for yourself, when your cravings for greens increases exponentially in the coming weeks on the Longevity

Paradox program. An additional benefit of greens is that they tend to suppress your appetite for the bad stuff that hurts our gut buddies. Studies have shown, for instance, that the phytochemicals in spinach dramatically reduce hunger for simple sugars and fats,[2] which is one reason I usually have the Green Smoothie in this book for breakfast.

Spinach is an ingredient in a lot of the greens blend powders on the market, but a word of warning about these phytochemical powders: I have not been able to find a greens blend without wheatgrass, barley grass, or oat grass as an ingredient—and the lectins in grains and grasses are the last things you need to swallow. I designed my own green formula called Primal Plants, combining spinach extract with eleven other superfood greens, particularly diindolylmethane (DIM), a remarkable immune-stimulating compound found in only minute amounts in broccoli. My blend also includes modified citrus pectin and galactooligosaccharides as hunger suppressants and gut buddy stimulators.

You can get the benefits without using this particular product, though. Spinach extract is available in 500-milligram capsules, and I recommend that you take two per day. DIM is available in capsule form, and the usual dose is 100 milligrams a day. Modified citrus pectin comes as a powder or in 500-milligram capsules.

Take two capsules or one scoop per day. My studies show that modified citrus pectin reduces an elevated galectin-3 level, a key marker of myocardial and kidney stress, by decreasing the bad bugs in your gut and increasing the good ones.

Prebiotics

The nomenclature of what goes on in your intestinal tract is confusing at best. *Pro*biotics, as you now know, refer to the bugs that live in and on you. *Pre*biotics are the compounds that the probiotics need to eat in order to survive and grow. It turns out that many of the compounds that are used for the treatment of constipation, such as psyllium powder or husks, work not as a bowel stimulator laxative but as a food for your gut buddies; this makes them grow and multiply, accounting for that bigger bowel movement. Even more interesting is the fact that the gang members in your gut can't eat psyllium husks and other fibers, so prebiotics feed the good guys and starve the bad guys.

One of the best prebiotics is inulin. I designed a practical way around the need to get these prebiotic fibers into you: PrebioThrive. It combines five prebiotics, including fructooligosaccharides (FOSs) and galactooligosaccharides (GOSs), in a powder that you simply mix with water and drink daily. Another way to get prebiotics is to take powdered psyllium husk. Start with a teaspoon a day in water and work up to a tablespoon a day. Also consider ordering GOSs, which are available online—I like the brand Bimuno. Take a packet or scoop each day. Then add a teaspoon of inulin powder a day.

Lectin Blockers

Despite our best efforts, we all sometimes find ourselves in situations in which we must—or accidentally—eat foods that contain major amounts of lectins. The good news is that there are a number of helpful lectin-absorbing compounds on the market. I designed a formula early in my career to help myself in such situations, and after many requests from my patients it is now available as Lectin Shield. It combines nine proven ingredients to absorb lectins or block them from reaching your gut wall. Simply take two capsules before a lectin-containing meal.

Alternatively, you could take glucosamine and methylsulfonylmethane (MSM) in tablet form, which also bind lectins. Products such as Osteo Bi-Flex and Move Free are available at Costco and other larger retailers. Also consider taking D-mannose, which is also in my Lectin Shield, in a dose of 500 milligrams twice a day, particularly if you are prone to urinary tract infections. D-mannose is the active ingredient in cranberries, but drinking the juice alone will not confer the same benefits, as D-mannose is present in cranberries in very small amounts.

Sugar Defense

Speaking of sugar: as you well know, we are awash in it—not only in its most familiar form but also in high-fructose corn syrup and any simple carbohydrate that rapidly breaks down into sugar, including your favorite fruits. I have been impressed through the years by the fact that the addition of a few simple supplements has made a major difference in my compliant patients' glucose and hemoglobin A1c (HbA1c) levels. And, as you remember, IGF-1 is influenced primarily by sugar and animal proteins, so the lower I can get your absorption and processing of sugar, the better. Last year, I formulated Glucose Defense. It combines chromium, zinc, selenium, cinnamon bark extract, berberine, turmeric extract, and black pepper extract. (The latter enhances the absorbability of turmeric—anytime you consume turmeric, make sure you also consume black pepper! Most high-quality turmeric supplements contain both spices.) You can take just two capsules twice a day to receive the whole spectrum of benefits.

If you prefer, Costco sells a wonderful product called CinSulin, which combines chromium and cinnamon.

Take two capsules a day. Combine this with 30 milligrams of zinc once a day, 150 micrograms of selenium a day, 250 milligrams of berberine twice a day, and 200 milligrams of turmeric extract a day.

Costco also offers an excellent turmeric supplement made by Youtheory. Take two of those a day. Because turmeric is so poorly absorbed, very little reaches your bloodstream. This is a shame because curcumin, the active ingredient in turmeric, is one of the few antioxidants to cross the blood-brain barrier into your brain. Because of this, I now produce Biomax Curcumin, which is absorbed via a different mechanism and thus reaches much higher blood levels; I currently take two daily.

While I'm on the subject of curcumin and berberine (sometimes called Oregon grape root, not to be confused with grape seed extract, a different polyphenol), both of these substances have been shown to act on PCSK9 gene expression in the liver. If that sounds familiar, the new injectable cholesterol-lowering drug Repatha works via the same mechanism (but at a cost of about $1,000 a month). Again, both berberine and curcumin supplements are easy to find.

Long-Chain Omega-3s

I have been measuring red blood cell–bound (RBC-bound) omega-3 levels in my patients for twelve years, and what I see scares me. Most people are profoundly deficient in the omega-3 fatty acids eicosapentaenoic acid (EPA) and, more important, docosahexaenoic acid (DHA). In fact, the only people in my practice who have sufficient levels of these brain-boosting fats without taking supplements eat sardines or herring on a daily basis. Why should you worry? Well, your brain is made up of approximately 60 percent fat. In other words, ladies, when you want to call your husband or boyfriend a "fathead," you unknowingly speak the truth! Half of the fat in your brain is DHA, and the other half is arachidonic acid (AA)—great sources of which are egg yolks and shellfish. Studies show that people with the highest levels of omega-3 fats in their blood have a better memory and a bigger brain than people with the lowest levels.[3] If that isn't persuasive enough, remember that fish oil helps repair your gut wall and keeps those nasty LPSs from getting across your gut border.

I recommend choosing a fish oil that is molecularly distilled. I've been so impressed with the longevity data coming from the tiny fishing village of Acciaroli, in southern Italy, where the diet is heavily based on anchovies and rosemary, that I've formulated my own omega-3 supplement with DHA and rosemary extract called Omega Advanced.

When taking fish oil, try to take 1,000 milligrams of

DHA per day. On the back of the bottle, you will find the serving size—either per capsule or per teaspoon if it is a liquid; then look under "ingredients" to find the DHA content per capsule or teaspoon. Calculate how many capsules or teaspoons will get you at or above 1,000 milligrams of DHA per day.

There are several good national brands. Nature's Bounty, 1400 milligrams, available at Costco, is the supplement I recommend currently. OmegaVia DHA 600 is a nice small capsule that's easy to swallow. And Carlson's Elite Gems can be swallowed or chewed. Carlson also makes an excellent lemon-flavored cod-liver oil. Take a tablespoon a day to get the right levels in your blood and brain. I've already mentioned that good-quality algae-derived DHA is now widely available for my vegan readers, but just make sure that you are taking the equivalent of 1,000 milligrams of DHA per day.

Mitochondrial Boosters

There are a number of compounds that I have discussed throughout the book that deserve our consideration in protecting and stimulating mitochondria. To name a few: N-acetyl L-cysteine (NAC), 500 milligrams; gynostemma extract, 450 milligrams; shilajit, 300 milligrams; reduced or L-glutathione, 150 milligrams; pau d'arco, 50 milligrams; pyrroloquinoline quinone (PQQ),

20 milligrams; and nicotinamide adenine dinucleotide, reduced (NADH), 10 milligrams. Feel free to add the doses listed above to your regimen.

Speaking of NADH, there are several compounds available that will probably boost your nicotinamide adenine dinucleotide, oxidized (NAD+) levels; one is nicotinamide riboside, patented and marketed as TRU Niagen. A recent human study suggested that a dose of 1,000 milligrams a day raises the NAD+ level in mononuclear cells. Not yet clinically available at a reasonable price is nicotinamide mononucleotide, which my friend and fellow longevity researcher David Sinclair of Harvard Medical School and MIT has shown to be more efficacious than Niagen in mouse trials. If cost is a concern, plain, cheap niacinamide may, in fact, have the same effect. Why raise your NAD+ level? To activate the SIRT1 gene, which in turn, suppresses mTOR. Suppress mTOR, and you will live better and longer. But remember, fasting has the exact same effect, and it's not only free, but actually saves money, because you don't pay for food!

Supplementation During Rapid Weight Loss or Fasting

Although I and other colleagues, including Dr. Jason Fung, MD, a leading expert on intermittent fasting, are proponents of intermittent, time-restricted, or water fasting, little attention has been given to the pioneering work of Dr. Roy Walford on the release of heavy metals and other toxins from our fat stores during this process. We store heavy metals and other organotoxins such as polychlorinated biphenyls (PCBs) and dioxin in our fat cells, where they remain remarkably inert (which explains why a swordfish with huge amounts of mercury in its flesh can happily swim around unimpaired). But as Dr. Walford found during the rapid weight loss that occurred in the failed Biosphere 2 experiment, his and the other crew members' heavy metal and other toxins blood levels rapidly climbed and remained severely elevated for about a year before returning to normal. This is because of our liver's and kidneys' very poor ability to have enough Phase 1 and 2 detoxification pathways to handle these toxins; moreover, many are excreted from our liver into our gut, where they are reabsorbed again.

Based on Dr Walford's findings, I recommend aiming to lose no more than 50 pounds a year, 25 pounds in six months, or 12.5 pounds in three months. But equally important, during fasting I recommend supplementing with milk thistle, D-limonene, dandelion, N-acetylcysteine, activated charcoal, and chlorella, the first four to activate the liver detoxification pathways and the last two to absorb the toxins and heavy metals re-excreted into the gut from the liver.

Other Supplements

Because so many of you have asked, I am listing here the supplements that I currently take. By no means does the fact that I take them mean that you should. Also, I change supplements often, based on new research or the results of my laboratory test values or those of my patients.

So here's the list in no particular order of importance.

Alpha GPC
Spirulina tablets
Cordyceps
Vitamin K2
Vitamin D3
Black cumin seed oil
Ashwagandha
Borage oil
Ginseng
Aged garlic extract
Alpha-tocopherol
Coriolus
Lion's mane
Chaga
Turkey tail
Reishi
Omega-7
(sea buckthorn oil)
Nutmeg
Hops extract
Apigenin
Fisetin
L-carnosine
Brown seaweed
Hyaluronic acid
Ubiquinol
Rosemary extract
Sage leaf extract

Alpha-carotene
Lutein
Luteolin complex with rutin
Artichoke extract
Quercetin
Pterostilbene
Vitamin C (timed release)
Trimethylglycine (TMG)
Naringin
Cloves
Saffron extract
Sesame seed lignans
HMR lignans
Pomegranate oil
Allithiamine
Benfotiamine
L-lysine
L-proline
Pregnenolone
R-alpha-lipoic acid
81 milligrams enteric-coated aspirin
Agmatine
Triphala
PQQ
L-glutathione
Myrosinase-activated sulforaphane glucosinolate (SGS)
Relora
Black raspberry
Moringa
Butyric acid
Ginkgo biloba
Coffee fruit extract
Diethylaminoethanol (DMAE)
Pyridoxal-5-phosphate (P-5-P)
Riboflavin-5-phosphate (R-5-P)
Biotin
BioSil (collagen, keratin, and elastin)
Chlorella
Mulberry extract
Glycine
Silk protein complex
N-A-C Sustain (N-acetyl L-cysteine)
Lithium

Potassium-magnesium aspartate
Glucomannan
Fenugreek
Thyme
Camu camu
Nicotinamide riboside
Nicotinamide mononucleotide
Cranberry seed oil
Parsley capsules

Yes, that's a lot, but as I said, you absolutely do not have to take all of them! To start, I recommend simply taking vitamin D, the B vitamins, and the G7. But even taking no supplements at all, following the Longevity Paradox program will give you plenty of benefits. I can't wait to hear from you about them.

Chapter 12
Longevity Paradox Recipes

You've already seen the food lists and meal plans, so hopefully you have a good idea of what types of foods you'll be eating on the Longevity Paradox program—a delicious variety of meals that will nourish your gut buddies and drive away the bad bugs. I have developed each of these recipes to support the longevity of your body and its inhabitants. That means plenty of prebiotic fiber, polyphenols, olive oil, spermidine, and all of the other wonderful foods your gut buddies love most. So dig in and enjoy—it's time to start cooking and eating for the 99 percent!

Soups and Salads

Longevity Leek Soup

The leeks in this soup are a wonderful longevity food with loads of polyphenols. Even better, it is perfect to eat during your five-day "fast." It has a bright lemony flavor with a richness from the nutmeg that will keep you warm all day.

Serves 4 to 6

2 tablespoons extra-virgin olive oil
1 pound leeks, cleaned and chopped
2 stalks celery, diced
3 cloves garlic, minced
1 tablespoon chopped fresh thyme
Zest of 1 lemon
1 large head cauliflower, cut into 2-inch florets
½ teaspoon fresh nutmeg
1 teaspoon fine sea salt, or more to taste
2 teaspoons coarse black pepper
2 quarts vegetable stock
1 bay leaf
Finely chopped chives for garnish

In a large soup pot, heat the olive oil over medium-high heat. Add the leeks, celery, garlic, thyme, lemon zest, and cauliflower along with the nutmeg, salt, and pepper, and sauté over medium heat, stirring regularly until the leeks begin to wilt.

Add the stock and the bay leaf and cook, covered, for 25 to 35 minutes, until the cauliflower is very tender.

Blend using a stick blender, or transfer into a regular blender and blend until smooth (work in batches so as not to overfill the blender).

Once pureed, return to the heat and cook for an additional 10 to 15 minutes. Taste, and adjust seasoning as needed.

Serve garnished with chopped chives.

Lentil Miso Soup with Shiitake Mushrooms

When the weather's chilly, there's nothing like a rich, earthy bean soup—and this one is full of polyamines and other antiaging compounds, and free of lectins!

Serves 4

2 tablespoons extra-virgin olive oil
1 large shallot, finely chopped
3 cloves garlic, minced

(continued)

1 cup thinly sliced fresh shiitake mushrooms
1½ tablespoons fresh thyme, minced
1 tablespoon fresh rosemary, minced
3 tablespoons red miso paste
6 cups Parmesan "Bone" Broth (page 430) or
mushroom broth
1½ cups pressure-cooked lentils
(Eden brand canned lentils okay)
1 cup sliced kale, stems removed
Coconut aminos, to taste

In a large soup pot, heat the olive oil over medium-high heat. Add the shallot and garlic and cook, stirring frequently, until the shallot is tender and the garlic is fragrant, about 3 minutes.

Reduce the heat to medium and add the mushrooms, thyme, and rosemary. Cook an additional 3 to 4 minutes, stirring frequently, until the mushrooms are tender.

Add the miso paste and cook, stirring constantly, until the paste is incorporated into the vegetable mixture.

Add the broth and lentils and cook for 20 to 30 minutes, covered.

Add the kale and cook, uncovered, an additional 20 minutes, until the kale is wilted and the soup is slightly thickened.

Add the coconut aminos a little bit at a time, tasting until you like the flavor, then serve.

Creamy Cauliflower Parmesan Soup

This soup is best made with the Parmesan "Bone" Broth on page 430—it really highlights the flavors of the cauliflower. If you love a leek and potato soup or chowder, chances are this soup will be right up your alley. Plus, it's full of cruciferous cauliflower and brain-boosting olive oil.

Serves 6

3 tablespoons extra-virgin olive oil
1 sweet onion, minced
2 stalks celery, diced
3 cloves garlic, minced
2 large heads cauliflower, cut into 2-inch florets
½ teaspoon fresh ground nutmeg
1 teaspoon fine sea salt, or to taste
2 teaspoons coarse black pepper
1 tablespoon white miso paste
7 cups mushroom stock or Parmesan "Bone" Broth (page 430)

(continued)

2 cups coconut milk

¼ cup grated Parmesan cheese or nutritional yeast

1 bay leaf

Finely chopped chives or thyme for garnish

In a large soup pot, heat the olive oil over medium-high heat. Add the onion, celery, garlic, and cauliflower, along with the nutmeg, salt, and pepper, and sauté over medium heat, stirring regularly, until the leeks begin to wilt.

Add the miso paste and cook, stirring, until the paste is incorporated.

Add the stock, coconut milk, Parmesan, and bay leaf and cook, covered, for 35 to 45 minutes, until the cauliflower is very tender.

Blend using a stick blender, or transfer into a regular blender and blend until smooth (work in batches so as not to overfill the blender).

Once pureed, return to the heat and cook an additional 10 to 15 minutes. If it is too thick, thin with a little water.

Taste and adjust the seasoning as needed.

Serve garnished with chopped herbs and additional grated Parmesan.

Bitter Green Salad with Walnut Blue Cheese Dressing

As I always say, more bitter, more better. Your favorite gut buddy, Akkermansia, loves these greens! But have no fear if you're not a fan of super bitter flavors and still want to benefit from consuming bitter foods. The fat in this salad dressing and the sweetness of the cranberries balance out the bitterness from the veggies beautifully.

Serves 2

FOR THE DRESSING:

¼ cup crumbled aged blue cheese, preferably French or Italian
¼ cup red wine vinegar
¼ cup extra-virgin olive oil
½ cup toasted walnuts
Juice of ½ lemon

FOR THE SALAD:

2 cups shredded kale
1 cup shredded or chopped endive or radicchio
¼ cup minced fresh dill (I admit I'd omit this, but my wife loves it)
¼ cup minced fresh parsley

(continued)

1 avocado, cut into chunks

¼ cup unsweetened dried cranberries

MAKE THE DRESSING: Combine all the dressing ingredients in a blender or in a food processor fitted with an S blade.

Pulse until smooth, thinning with water as needed (it should be the consistency of ranch or blue cheese dressing).

MAKE THE SALAD: Combine the kale, endive, dill, and parsley in a large bowl.

Add half the dressing and toss until the greens are well coated.

Top the salad with avocado and cranberries, and serve with the remaining dressing.

Arugula Salad with Hemp Tofu, Grain-Free Tempeh, or Cauliflower "Steak" and Lemon Vinaigrette

This is another great option for your five-day "fast" that you can put together easily to take to work for lunch or throw together for dinner at the end of a long day.

Serves 1

FOR THE TEMPEH:

1 tablespoon avocado oil

4 grain-free tempeh, cut into ½-inch-thick strips*

1 tablespoon freshly squeezed lemon juice

¼ teaspoon sea salt, preferably iodized

FOR THE DRESSING:

2 tablespoons extra-virgin olive oil

1 tablespoon freshly squeezed lemon juice

Pinch of sea salt, preferably iodized

FOR THE SALAD:

1½ cups arugula

Zest of ½ lemon (optional)

MAKE THE TEMPEH: In a small skillet, heat the avocado oil over medium heat. Place the tempeh strips in the

(continued)

hot pan and sprinkle them with the lemon juice and salt. Sauté the tempeh strips for about 2 minutes; turn them and sauté for another 2 minutes, until cooked through. Remove from the pan and reserve.

MAKE THE DRESSING: Combine all the dressing ingredients in a mason jar with a tight-fitting lid (double the ingredients if making two batches). Shake until well combined.

MAKE THE SALAD: Toss the arugula in the dressing and top with the tofu, tempeh, or cauliflower steak, adding the lemon zest, if desired.

OTHER VEGETARIAN VERSIONS: In place of the tempeh, tofu, or cauliflower, substitute acceptable Quorn products: Chik'n Tenders, Grounds, Turk'y Roast, Chik'n Cutlets. (They contain a tiny amount of egg white so are not totally animal protein free, but the amount is probably negligible in terms of mTOR.)

*You can replace the tempeh with hemp tofu or a cauliflower "steak" (a ¾-inch-thick cauliflower slice seared over high heat in avocado oil until golden brown on both sides).

Romaine Salad with Avocado, Cilantro Pesto, and Grain-Free Tempeh

This satisfying salad will keep you full and energized on the five-day "fast." To save time, make the cilantro pesto in advance and store for up to three days in the refrigerator in a covered glass container. You can substitute basil or parsley for the cilantro.

Serves 1

FOR THE TEMPEH:

1 tablespoon avocado oil
4 grain-free tempeh, cut into ½-inch-thick strips*
1 tablespoon freshly squeezed lemon juice
¼ teaspoon sea salt, preferably iodized

FOR THE PESTO:

2 cups chopped cilantro (if you, like me, taste "soap" when you eat cilantro, substitute Italian parsley)
¼ cup extra-virgin olive oil
2 tablespoons freshly squeezed lemon juice
¼ teaspoon sea salt, preferably iodized

(continued)

FOR THE DRESSING:

½ avocado, diced

2 tablespoons freshly squeezed lemon juice

2 tablespoons first-pressed extra-virgin olive oil

Pinch of sea salt, preferably iodized

FOR THE SALAD:

1½ cups chopped romaine lettuce

MAKE THE TEMPEH: In a small skillet, heat the avocado oil over medium heat. Place the tempeh strips in the hot pan and sprinkle them with the lemon juice and salt. Sauté the tempeh strips for about 2 minutes; turn them and sauté for another 2 minutes, until cooked through. Remove from the pan and reserve.

MAKE THE PESTO: Place all the pesto ingredients in a high-powered blender. Process on high until very smooth.

MAKE THE DRESSING: Toss the avocado in 1 tablespoon of the lemon juice and set aside. Combine the remaining 1 tablespoon lemon juice, the olive oil, and the salt in a mason jar with a tight-fitting lid. (Double the ingredients if making two batches.) Shake until well combined.

MAKE THE SALAD: Toss the romaine in the dressing. Arrange the avocado and tempeh over the lettuce and spread the pesto on top.

*You can replace the tempeh with hemp tofu or a cauliflower "steak" (a ¾-inch-thick cauliflower slice seared over high heat in avocado oil until golden brown on both sides).

Hemp Tofu-Arugula-Avocado Seaweed Wrap with Cilantro Dipping Sauce

Nori is a form of seaweed that has been flattened into squares or strips. It makes a terrific stand-in for flatbread in this wrap, which can be eaten as part of your five-day "fast."

Serves 1

FOR THE FILLING:

1 tablespoon avocado oil
4 ounces hemp tofu, cut into ½-inch-thick strips
2 tablespoons freshly squeezed lemon juice
¼ teaspoon sea salt, preferably iodized, plus additional to taste
½ avocado, diced

(continued)

FOR THE DIPPING SAUCE:

2 cups chopped fresh cilantro

¼ cup first-pressed extra-virgin olive oil

2 tablespoons freshly squeezed lemon juice

¼ teaspoon sea salt, preferably iodized

FOR THE WRAPS:

1 cup arugula

1 sheet nori (sushi seaweed)

4 green olives, pitted and halved

Sea salt, to taste

MAKE THE FILLING: In a small skillet, heat the avocado oil over medium-high heat. Place the hemp tofu strips in the hot pan and sprinkle with 1 tablespoon of the lemon juice and the salt. Sauté the strips for about 2 minutes; turn them and sauté for another 2 minutes, until cooked through. Remove from the pan and reserve.

Toss the avocado in the remaining tablespoon lemon juice and season with salt. Set aside.

MAKE THE DIPPING SAUCE: Place all the dipping sauce ingredients in a high-powered blender. Process on high until very smooth.

MAKE THE WRAPS*: Spread the arugula over the bottom half of the seaweed sheet. Top with the filling and olives. Sprinkle with salt to taste. Carefully roll into a tight wrap, sealing the end with a little water. Cut in half and serve with the cilantro dipping sauce.

OTHER VEGETARIAN VERSIONS: In place of the tempeh, tofu, or cauliflower, substitute acceptable Quorn products: Chik'n Tenders, Grounds, Turk'y Roast, Chik'n Cutlets. (They contain a tiny amount of egg white so are not totally animal protein free, but the amount is probably negligible in terms of mTOR.)

OTHER VEGAN VERSIONS: Replace the hemp tofu with grain-free tempeh or a cauliflower "steak" (a ¾-inch-thick cauliflower slice seared over high heat in avocado oil until golden brown on both sides).

*A bamboo mat, available in the Asian foods section of most supermarkets, can help you roll tight seaweed wraps.

Romaine Lettuce Boats Filled with Guacamole

I recommend that you use Hass avocados for your guacamole (and other recipes). Hass have a black or dark green pebbly skin and contain more fat (the heart-healthy monounsaturated kind) than do the larger, smooth-skinned Florida avocados, which tend to be more watery.

Serves 1

½ avocado
1 tablespoon finely chopped red onion
1 teaspoon finely chopped fresh cilantro
1 tablespoon freshly squeezed lemon juice
Pinch of sea salt, preferably iodized
4 large romaine lettuce leaves, washed and patted dry

Place the avocado, onion, cilantro, lemon juice, and salt in a bowl and mash with a fork until smooth.

To serve, scoop an equal amount of the guacamole into each lettuce leaf.

Lemony Brussels Sprouts, Kale, and Onions with Cabbage "Steak"

You can use any of the many types of kale in this hearty vegetable dish. Unless you're using baby kale, remove the stems before chopping. (There is no need to remove the stems of or chop baby kale.)

Serves 1

4 tablespoons avocado oil
One 1-inch-thick red cabbage slice
¼ teaspoon plus 1 pinch of sea salt, preferably iodized
½ red onion, thinly sliced
1 cup Brussels sprouts, thinly sliced
1½ cups chopped kale
1 tablespoon freshly squeezed lemon juice
First-pressed extra-virgin olive oil (optional)

Heat a skillet over high heat. When it is hot, add 1 tablespoon of the avocado oil, reduce the heat to medium, and sear the cabbage slice until it is golden brown on one side, about 3 minutes. Flip it and brown it on the other side. Season with the pinch of salt, remove to a plate, and cover to keep warm. Wipe the skillet clean with a paper towel and return it to the stovetop.

(continued)

Heat 2 tablespoons of the avocado oil in the skillet over medium heat. Add the onion and Brussels sprouts. Sauté until tender, about 3 minutes. Add the remaining 1 tablespoon avocado oil, the kale, and the lemon juice and sauté for another 3 minutes, until the kale is wilted. Season with the ¼ teaspoon salt.

To serve, top the cabbage "steak" with the sautéed vegetables. Add a drizzle of olive oil, if desired.

Cabbage-Kale Sauté with Grain-Free Tempeh and Avocado

This tasty dish makes a great substitute for a grain bowl and is very adaptable. Be sure to use bok choy or Napa cabbage instead of green cabbage if eating this during your five-day "fast."

Serves 1

½ avocado, diced
3 tablespoons freshly squeezed lemon juice
4 pinches of sea salt, preferably iodized
3 tablespoons avocado oil
1½ cups thinly sliced green cabbage
½ red onion, thinly sliced
4 ounces grain-free tempeh

Toss the avocado in 1 tablespoon of the lemon juice and season with a pinch of salt. Set aside.

Heat a skillet over medium heat. When it is hot, add 2 tablespoons of the avocado oil and the cabbage and onion. Sauté until tender, about 10 minutes, stirring occasionally. Season with 2 more pinches of salt. Using a slotted spatula, remove from the skillet and set aside.

Add the remaining 1 tablespoon avocado oil to the skillet, raise the heat to high, and add the remaining 2 tablespoons lemon juice and the tempeh. Sear the tempeh, flipping after 3 minutes, until cooked through, about 6 minutes total. Season with the remaining pinch of salt.

To serve, top the sautéed cabbage and onions with the tempeh and avocado.

OTHER VEGETARIAN VERSIONS: In place of the tempeh, tofu, or cauliflower, substitute acceptable Quorn products: Chik'n Tenders, Grounds, Turk'y Roast, Chik'n Cutlets. (They contain a tiny amount of egg white so are not totally animal protein free, but the amount is probably negligible in terms of mTOR.)

OTHER VEGAN OPTIONS: Replace the grain-free tempeh with hemp tofu or a cauliflower "steak" (a ¾-inch-thick cauliflower slice seared over high heat in avocado oil until golden brown on both sides).

Roasted Broccoli with Cauliflower "Rice" and Sautéed Onions

I love a good veggie stir fry! To make cauliflower "rice," grate the cauliflower with a cheese grater, using the largest holes, into rice-shaped pieces. You can also pulse it in a food processor, using the S blade and being careful not to overprocess it. If you use a food processor, cut the cauliflower into chunks first.

Serves 1

1½ cups broccoli florets
2½ tablespoons avocado oil
3 pinches of sea salt, preferably iodized
½ head medium cauliflower, riced
1 tablespoon freshly squeezed lemon juice
¼ teaspoon curry powder
½ red onion, thinly sliced

Preheat the oven to 375° F.

Put the broccoli into a Pyrex dish with 1 tablespoon of the avocado oil. Roast in the oven for 15 minutes, stirring frequently, until tender. Season with a pinch of salt.

Sauté the cauliflower in a medium skillet with 1 tablespoon of the avocado oil, the lemon juice, the curry powder, and a pinch of salt until tender, 3 to 5 minutes.

Do not let it get mushy by overcooking. Transfer the cauliflower "rice" to a plate and wipe the skillet clean with a paper towel.

Reheat the skillet over medium heat. When it is hot, add the remaining ½ tablespoon avocado oil and the sliced onion and sauté until tender, stirring frequently, for about 5 minutes. Season with a pinch of salt.

To serve, place the cauliflower "rice" on a plate and top with the broccoli and sautéed onions.

Green Mango and Pear Salad

This tangy, refreshing dish was inspired by the green mango salad served in my local Thai restaurant. The unripe mangoes are favorites of your gut buddies, and the cabbage has loads of prebiotic fiber. Feel free to add a peeled, seeded chili pepper for some heat if you prefer spicy food.

Serves 4

¼ cup fish sauce or coconut aminos or one-half amount of each to taste
Juice and zest of 2 limes
1 tablespoon unsweetened coconut milk
2 tablespoons toasted sesame oil

(continued)

1 packet stevia or 1 tablespoon yacon syrup
1 small red onion, thinly sliced
2 large unripe mangoes, peels and pits removed, shredded
1 cup shredded cabbage
1 large green pear, peeled and shredded
1 shredded carrot
¼ cup chopped fresh cilantro
¼ cup chopped macadamia nuts

In a large bowl, combine the fish sauce, lime juice and zest, coconut milk, sesame oil, and stevia.

Add the onion, mangoes, cabbage, pear, and carrot, and toss to combine.

Top with cilantro and macadamia nuts before serving.

Spinach Salad with Lentil-Cauliflower Fritters

This salad is incredibly fresh tasting, thanks to plenty of mint leaves mixed in with the greens. Think of the fritters on top as the most flavorful croutons you'll ever try—nutty, a little creamy, and a little cheesy—a perfect topping for any salad and full of prebiotic fiber for your gut buddies to enjoy along with you.

Serves 2

FOR THE FRITTERS:

1 cup pressure-cooked lentils

1 cup cauliflower rice

¼ cup parsley

1 tablespoon sesame tahini

½ teaspoon sea salt

¼ teaspoon black pepper

¼ teaspoon garlic powder

¼ teaspoon paprika

¼ cup Parmesan cheese or nutritional yeast

1 egg

2 tablespoons cassava flour plus more as needed

¼ cup extra-virgin olive oil

FOR THE SALAD:

¼ cup balsamic vinegar

¼ cup extra-virgin olive oil

1 tablespoon Dijon mustard

¼ teaspoon sea salt

½ cup minced fresh mint

6 cups baby spinach, rinsed and dried

1 cup packaged broccoli slaw

MAKE THE FRITTERS: Add the lentils, cauliflower rice, and parsley to a food processor fitted with an S blade and pulse until well combined. *(continued)*

Add the sesame tahini, salt, pepper, garlic powder, paprika, Parmesan, egg, and 2 tablespoons cassava flour and blend an additional minute, until smooth.

Pinch a small bit of the mixture between your fingers. If it holds its shape and doesn't feel wet, you're good to cook. If not, add additional cassava flour, 1 teaspoon at a time, until fritters form easily.

Let the fritter mixture rest while you heat the oil in a large skillet. When the oil is hot and shimmering in the pan, drop tablespoon-sized balls of the batter into the oil.

Cook 3 to 4 minutes per side, then let rest on a paper towel while assembling the salad.

MAKE THE SALAD: Combine the vinegar, oil, mustard, salt, and mint in a large salad bowl.

Add the spinach and broccoli slaw and toss until well dressed.

Serve topped with the fritters.

Tangy Sesame Slaw

There's nothing like coleslaw for a summer barbecue side dish—and this creamy, rich slaw is one of my favorites, because it's not heavy and gloppy with mayonnaise. Instead, it features nutty tahini and avocado to provide the creaminess, plus plenty of lemon juice for tang.

Serves 4

¼ cup sesame tahini
Juice of 1 lemon
1 clove garlic, crushed
1 tablespoon sesame oil
1 tablespoon yacon syrup
1 ripe avocado, mashed
1 tablespoon coconut aminos
1 large red onion, thinly sliced
1 small head cabbage, shredded
1 beet, shredded
1 carrot, shredded
¼ cup finely chopped fresh mint
¼ cup finely chopped fresh dill

In a large bowl, whisk together the tahini, lemon juice, garlic, sesame oil, and yacon syrup until smooth.

(continued)

Add the avocado and coconut aminos and continue whisking until a very smooth, thick dressing the consistency of mayonnaise is formed.

Add the remaining ingredients to the bowl and toss until well combined.

Serve as is for a side dish, or top with a couple of fried omega-3 eggs to make a full meal.

Entrees

Sweet Potato Gnocchi with Creamy Mushroom Sauce

Classic sweet potato, nutmeg, and sage flavors taste like fall to me, but that's no reason you can't enjoy this rich, comforting dish year-round, or at least whenever you want to indulge yourself the healthy way. Sweet potatoes are a wonderful source of resistant starch, much like the tubers that help naked mole rats live such long and healthy lives.

Serves 4

FOR THE GNOCCHI:

1 pound peeled sweet potato or yam, cut into large chunks (about 1 large sweet potato)

1 large omega-3 egg or egg substitute, such as Bob's Red Mill Vegetarian Egg Replacer

1½ cups cassava flour (do not use tapioca flour as a substitute)

(continued)

½ teaspoon sea salt
½ teaspoon grated fresh nutmeg

FOR THE SAUCE:

1 tablespoon coconut oil or grass-fed European butter
12 ounces mushrooms (shiitake, portabella, cremini, oyster, or white button), diced
1 clove garlic, minced
1 teaspoon minced fresh thyme
1¼ cups unsweetened coconut milk
Juice and zest of ½ lemon
1 tablespoon minced fresh parsley
¼ cup grated Parmesan cheese or nutritional yeast
½ teaspoon iodized sea salt or more to taste
½ teaspoon fresh black pepper

MAKE THE GNOCCHI: Place the sweet potatoes in a large pot and cover with water. Bring to a boil, then reduce to a simmer. Cover and cook 15 to 20 minutes, or until tender.

Remove from heat and let cool to room temperature, then drain, transfer to a large bowl, and mash with a potato masher until smooth.

Double-check that the sweet potato mash is cool (so you don't cook the egg), then add the egg, 1 cup of the cassava flour, and the salt and nutmeg to the mixture.

With clean hands, knead until a smooth dough is formed, adding the remaining flour as needed to form a dough that's neither sticky nor crumbly.

Bring a large pot of salted water to a boil.

While waiting for the water to boil, roll chunks of dough into long snakes, about the width of your thumb. Cut each snake into 1-inch pieces (about the length from the tip of your thumb to the first knuckle). Shape the gnocchi by either rolling each piece down the back of a fork or using your thumb to make a shallow indentation in each piece.

When the water boils, slide the gnocchi one by one into the water using a slotted spoon. When they float to the surface, remove them with the slotted spoon and store in a covered dish to keep warm.

If you have more gnocchi than you intend to serve immediately, spread them on a parchment-lined sheet pan once cooked. Allow to cool before transferring to a freezer. Freeze on the sheet pan; then, when solid, transfer to a zip-top bag to store.

MAKE THE SAUCE: In a large saucepan, heat the oil or butter over medium-high heat.

Add the mushrooms and cook, stirring frequently, for 3 to 5 minutes, or until the mushrooms are fragrant and tender.

Add the garlic and thyme and cook an additional minute, until the garlic is tender. *(continued)*

Pour the coconut milk into the pan along with the lemon juice and zest and cook, stirring frequently, until the coconut milk thickens, about 8 to 10 minutes.

Add the parsley, Parmesan, salt, and pepper. Stir until the cheese melts, then transfer the gnocchi to the sauce.

Cook an additional 2 to 3 minutes before serving.

Walnut Lentil Veggie Burgers or Meatballs

This veggie burger actually tastes meaty, thanks to the lentils and mushrooms. Plus, the walnuts add a rich kick of protein and cancer-fighting compounds, and the herbs add a bit of freshness. I love to make a "protein-style" burger with one of these patties, lettuce, avocado mayo, and a grilled onion wrapped in lettuce leaves.

Serves 4

½ red onion, coarsely chopped
1 clove garlic
½ cup walnuts
½ cup fresh shiitake or cremini mushrooms
¼ cup fresh parsley leaves

¾ teaspoon ground cumin
¾ teaspoon sweet paprika
½ teaspoon curry powder
½ teaspoon black pepper
½ teaspoon dried mustard powder
½ teaspoon sea salt
2 cups pressure-cooked lentils
(Eden brand canned lentils okay)
1 omega-3 egg or 1 vegan egg substitute
1 tablespoon ground flaxseed
¼ to ½ cup cassava flour

Preheat the oven to 350° F. Line a sheet pan with parchment and set aside.

In a food processor fitted with an S blade, pulse together the onion, garlic, walnuts, mushrooms, parsley, cumin, paprika, curry, black pepper, mustard powder, and sea salt until a smooth paste is formed (50 to 100 short pulses). Transfer to a bowl and fold in the lentils, egg, and flaxseed, crushing the lentils with your spoon or spatula as you fold them in.

Add 2 tablespoons cassava flour and let the mixture rest 5 minutes to absorb the liquid. With your fingers, test to see if the mixture forms a cohesive ball. Add flour bit by bit until the mixture holds its shape when molded.

(continued)

Form into 4 large patties or about 20 meatballs and space evenly on the sheet pan.

FOR THE PATTIES: Bake for 15 to 20 minutes, then carefully flip and bake for an additional 10 minutes.

FOR THE MEATBALLS: Bake for 20 minutes total, flipping every 5 minutes.

Serve on a bed of lettuce, over Miracle Noodle brand pasta, or with the Sweet Potato Gnocchi with Creamy Mushroom Sauce (page 401).

Roasted Broccoli with Miso Walnut Sauce

Love broccoli with cheddar cheese? Try this unexpected twist with lots of prebiotic fiber for your gut buddies. This side dish has savory notes similar to those of cheese but with a touch of added sweetness for great bold flavor.

Serves 4

½ cup walnuts, soaked in water at least 8 hours

⅓ cup red or white miso paste

1 tablespoon yacon syrup or local honey

4 tablespoons coconut aminos

¼ cup sesame oil
2 shallots, thinly sliced
5 cloves garlic, thinly sliced
5 cups broccoli florets

Preheat the oven to 400° F.

Drain the walnuts and pat dry with a kitchen towel.

In a food processor fitted with an S blade, combine the walnuts, miso paste, yacon syrup, coconut aminos, and sesame oil. Pulse until a thick paste is formed.

Transfer the paste into a large bowl along with the shallots, garlic, and broccoli and toss to coat the broccoli with the walnut mixture.

Transfer to a sheet pan and roast 15 minutes; flip the broccoli and continue to roast it until golden brown (10 to 15 additional minutes).

Serve hot or at room temperature.

Mushroom and Thyme Braised Tempeh

I love braised beef with mushrooms, but these days, I'm not much of a meat eater—that's where this braised tempeh comes in. This dish is tasty over a baked sweet potato, cauliflower rice, or cooked millet—or tossed with Miracle Noodle brand pasta (like beef Stroganoff).

Serves 4

¼ cup extra-virgin olive oil
2 8-ounce packages of plain tempeh, each cut into 8 to 10 slices
2 large shallots, minced, or 1 small red onion, minced
4 cups sliced crimini or portabella mushrooms
2 tablespoons minced fresh thyme
2 cloves garlic, minced
½ cup dry red wine (nice enough to drink)
¼ cup Dijon mustard
2 cups mushroom stock, Parmesan "Bone" Broth (page 430), or homemade beef stock
1 tablespoon arrowroot powder
¼ cup water
Iodized sea salt and black pepper, to taste

In a large pan, heat the oil over medium-high heat.

When the oil is hot, add the tempeh and sear 2 to 3 minutes per side, until golden brown. Remove from the oil and set aside.

Add the shallots and mushrooms to the pan and sauté, stirring frequently, until the mushrooms are tender and golden brown, about 5 minutes.

Add the thyme and garlic and cook an additional minute, until very fragrant.

Add the red wine and deglaze, scraping the bottom of the pan to remove any cooked-on bits.

Add the mustard and stir until well combined, then add the broth.

While broth is simmering, whisk together the arrowroot powder and water. Add to the broth mixture and stir to combine, then add the tempeh.

Reduce heat to low and simmer 20 to 30 minutes, until the sauce is very thick.

Season with salt and pepper (about ½ teaspoon each) before serving.

Spiced Refried "Beans" Made with Lentils

Love refried beans but finally convinced to give up lectins? Try these pressure-cooked refried lentils—they've got all the same classic spices as refried beans, plus that delicious silky texture, but they're actually good for you and your gut buddies, too. You can pressure-cook the lentils yourself or use Eden canned lentils to make your life easier.

Serves 4

3 cups pressure-cooked lentils
(Eden brand canned lentils are great)
1½ tablespoons extra-virgin olive oil
1 medium onion, finely minced
2 cloves garlic, finely minced
1 teaspoon cumin
1 teaspoon paprika
1 teaspoon garlic powder
1 teaspoon black pepper
½ teaspoon dried oregano
½ teaspoon dried sage
1 tablespoon coconut aminos
Juice of ½ lime
Cilantro, for garnish

Drain the cooked lentils well and pat dry.

In a large saucepan, heat the olive oil over medium-high heat.

Add the onion, garlic, cumin, paprika, garlic powder, pepper, oregano, and sage and cook 3 to 5 minutes, stirring frequently, until the onions and garlic are very tender and fragrant.

Add the lentils and cook, crushing the lentils with a spoon or spatula.

When the lentils are crushed smooth, stir in the coconut aminos and lime juice.

Serve garnished with cilantro.

Ginger Coconut Cauliflower "Rice"

This aromatic variation on fried rice is the perfect base for curries, grilled seafood, or even roasted vegetables. It's creamy but still manages to be nice and light, so it doesn't overpower a meal while providing lots of prebiotic fiber and ketones from coconut oil.

Serves 4

4 tablespoons coconut oil
1 small shallot, minced

(continued)

1 tablespoon minced ginger
4 cups cauliflower rice
¼ teaspoon iodized sea salt
1 cup coconut milk
1 cup shredded unsweetened coconut
Zest of 1 lime
1 tablespoon coconut aminos

In a large skillet, heat the coconut oil over medium-high heat.

Add the shallot and ginger and cook, stirring frequently, until fragrant.

Add the cauliflower rice, salt, coconut milk, and coconut and cook, stirring frequently, until the cauliflower rice is tender and the mixture is creamy.

Season with lime zest and coconut aminos before serving.

Lentil Broccoli Curry

Similar to a traditional Indian curry, this creamy, richly spiced red lentil curry features a new twist on classic flavors, thanks to ingredients such as broccoli rice. Try it over rice or boiled millet, or even serve it over a baked sweet potato.

Serves 4

¼ cup coconut oil
1 onion, finely minced
1 cup broccoli rice*
1 teaspoon ground cumin
1 tablespoon turmeric
1 teaspoon black pepper
½ teaspoon paprika
½ teaspoon sea salt
½ teaspoon mustard seeds
2 tablespoons curry powder
4 cloves garlic, minced
3 cups pressure-cooked red lentils (Eden brand canned lentils okay)
2 cups unsweetened coconut milk
Juice of 1 lemon

In a large pan, heat the oil over medium-high heat.

Add the onion, broccoli rice, cumin, turmeric, pepper, paprika, salt, mustard seeds, and curry powder. Cook, stirring frequently, until the onion is tender and the mixture is very fragrant.

Add the garlic and lentils and cook an additional 5 minutes, stirring frequently so that the garlic doesn't burn.

(continued)

Add the coconut milk and lemon juice and reduce heat to low.

Let simmer 20 to 30 minutes, until the mixture is very thick.

Serve on its own or over Miracle Rice.

*I buy it at Trader Joe's; you can also make it by pulsing broccoli stems in a food processor fitted with an S blade until rice-sized grains are left.

Toasted Millet "Grits" with Spicy Eggs

I used to love grits for breakfast when I lived in Georgia—and luckily, millet has a similar texture while being rich in magnesium and potassium, high in fiber, and free of lectins. For a savory breakfast, try simple toasted millet "grits" with mushrooms and eggs for a filling, hearty meal. It's also great at dinner—and in my opinion especially delicious when made with Parmesan "Bone" Broth.

Serves 2

2 tablespoons extra-virgin olive oil
1 shallot, minced
1 cup uncooked millet

2 cups mushroom broth or Parmesan "Bone" Broth (page 430)
½ teaspoon iodized sea salt
½ cup minced mushrooms
1 tablespoon minced fresh thyme
4 omega-3 eggs
1 teaspoon cayenne pepper (less if you *really* don't like spicy food)

In a large saucepan, heat 1 tablespoon of the oil over medium-high heat.

Add the shallot and millet and cook, stirring frequently, until the shallot is translucent and the millet smells "toasty."

Add the broth and salt and bring to a boil, then reduce heat to low. Cook, covered, until the millet is tender, 15 to 20 minutes.

While the millet is cooking, heat the remaining 1 tablespoon oil in a small skillet.

Add the mushrooms and thyme and cook, stirring occasionally, until the mushrooms are tender (about 3 minutes).

Add the eggs and cayenne pepper and cook, stirring frequently, until the eggs are scrambled.

Serve the eggs over the millet for a savory alternative to grits or oatmeal.

Mushroom Artichoke Fettuccini Bake

I used to love mushroom artichoke lasagna, but it's out of the question on the Longevity Paradox program—luckily, thanks to Miracle Noodle, I can get all the flavors (and texture) of that lasagna without the dangerous lectins and casein A1 dairy products. It's savory, creamy, a little salty, and full of meaty mushrooms with their cancer-fighting properties, spermidine, and earthy flavor.

Serves 4 to 6

Cooking spray
¼ cup avocado oil
1 large onion, minced
1 pound mushrooms, diced (portabella, cremini, shiitake, trumpet, or oyster are good)
1 pound artichoke hearts (frozen and thawed or canned and rinsed), roughly chopped
2 tablespoons minced fresh rosemary
2 tablespoons minced garlic
2 tablespoons minced fresh thyme
1 teaspoon sea salt
1 teaspoon black pepper
Zest of 1 lemon

3 packs of Miracle Noodle Fettuccini, prepared the Gundry way*
2 tablespoons cassava or coconut flour
2 cups coconut milk
1 cup mushroom broth or Parmesan "Bone" Broth (page 430)
½ cup Parmesan cheese or nutritional yeast
¼ cup finely chopped walnuts

Preheat the oven to 350° F. Spray a casserole dish with cooking spray and set aside.

In a large frying pan, heat the oil over medium-high heat.

Add the onion and mushrooms and cook, stirring frequently, until the onion is translucent and the mushrooms are tender.

Add the artichoke hearts, rosemary, garlic, thyme, salt, pepper, lemon zest, and fettuccini to the pan and cook an additional 2 to 3 minutes, until very fragrant.

Add the cassava flour and cook an additional minute, stirring until it is well combined with the mushroom mixture.

Add the coconut milk, the broth, and ¼ cup of the cheese and cook for 3 to 4 minutes, until it begins to thicken.

(continued)

Transfer the mixture to the prepared baking dish and top with the remaining cheese and the walnuts.

Bake 20 to 30 minutes, until golden brown and bubbly.

Let rest 5 to 10 minutes before serving.

*To do this, bring a pot of water to a boil. Rinse the pasta under cold water, then add to the boiling water. Let boil for 2 to 3 minutes, then rinse again under cold water for 2 minutes. Throw the pasta into a dry pan and cook off any remaining moisture over medium heat. Don't worry about the popping sounds; they're normal!

Cauliflower "Fried Rice"

This is a great, filling meal to enjoy during your five-day "fast." With options like this, you'll never go hungry even while your body thinks you're fasting. A win-win all around.

Serves 6 to 8

2 tablespoons sesame oil

1 medium yellow onion, diced

¼ cup minced green onions

1-inch piece ginger root, peeled and minced

2 cloves garlic, minced
1 cup thinly sliced mushrooms (any type)
4 ribs celery, thinly sliced
1 cup broccoli florets
4 ounces water chestnuts (canned okay), roughly chopped
4 cups cauliflower rice
1 tablespoon coconut aminos
¼ teaspoon paprika
¼ teaspoon powdered mustard

In a large skillet or wok, heat the oil over medium-high heat.

Add the onion, green onions, and ginger and cook several minutes, until the onions are translucent.

Add the garlic, mushrooms, celery, broccoli, and water chestnuts and cook, stirring frequently, until the vegetables soften and the garlic is fragrant (5 to 6 minutes).

Turn heat to high and add the cauliflower rice. Cook for 3 to 4 more minutes, stirring frequently to ensure that it doesn't burn.

After a minute, add the coconut aminos, paprika, and powdered mustard.

Continue cooking on high heat, stirring frequently, until the cauliflower is tender but not mushy, and serve.

Sweets

Sweet Potato and Coconut Pudding

This dessert is inspired by an Asian dessert featuring taro, but sweet potatoes are often easier to find in your grocery store and have lots of resistant starch in their own right. It's not the sweetest dessert, but the flavors are incredible: coconut, vanilla, and cinnamon play beautifully together, making it feel decadent and light at the same time.

Serves 4

1 cup tapioca pearls
2 cups coconut milk
1 cup unsweetened shredded coconut
¼ teaspoon cinnamon
1 teaspoon vanilla extract
¼ cup erythritol powder
2 cups peeled, diced sweet potatoes or taro root

In a small saucepan, bring 2 cups of water to a boil. Add the tapioca pearls and boil for 10 minutes, then remove from the heat and cover. Let rest for 20 minutes.

While the tapioca is cooking, heat the coconut milk, shredded coconut, cinnamon, vanilla, and erythritol powder over medium heat, stirring occasionally.

Add the sweet potatoes and continue to cook until tender, 15 to 20 minutes depending on how small you've cut them.

When the sweet potatoes are tender, strain the water off the tapioca pearls and add them to the coconut mixture. Cook for an additional 2 minutes.

Serve warm, or transfer to the refrigerator and serve chilled for a custardlike texture.

Blueberry Miso Muffins

Miso is delicious in desserts, thanks to the slight sweetness in white miso paste, which provides almost a butterscotch flavor—perfect with fresh, in-season fruit and pie-inspired seasoning.

Makes 12 muffins

¼ cup grass-fed butter or coconut oil
2 tablespoons white miso paste
2 large omega-3 eggs or Bob's Red Mill Vegetarian Egg Replacer
8 drops vanilla-flavored liquid stevia (or more to taste)
1 cup coconut milk
2 cups blanched almond flour
¼ cup coconut flour
1 teaspoon baking powder
¼ teaspoon allspice
¼ teaspoon nutmeg
¼ teaspoon cinnamon
½ cup fresh blueberries*

Preheat the oven to 350° F. Line a muffin tin with papers and set aside.

In a stand mixer fitted with a paddle attachment, beat the butter and miso paste on high until fluffy.

Add the eggs one at a time and beat until well blended.

Combine stevia and coconut milk in a measuring cup and set aside.

In a separate bowl, whisk together the almond flour, coconut flour, baking powder, allspice, nutmeg, and cinnamon.

Add half the dry ingredients to the egg mixture and beat to combine.

Add half the coconut milk mixture and beat to combine. Alternate adding wet and dry ingredients until the mixture is well blended.

Gently fold the blueberries into the batter (use a spatula, not the mixer), then scoop the batter into 12 muffin tins.

Bake 18 to 25 minutes, until a knife inserted into the center of each muffin comes out clean. Let cool before serving.

*If blueberries aren't in season, feel free to use ½ cup of in-season fruit, finely minced.

Mexican Chocolate "Rice" Pudding

I've always loved the sweet-spicy flavor profile of Mexican chocolate, which contains cinnamon and cayenne pepper, so I added that classic set of spices to my "rice" pudding—made with cauliflower rice for loads of prebiotic fiber. You won't believe that there's actually cauliflower in this dessert; it just tastes like rich, creamy, crave-worthy chocolate.

Serves 4

4 cups cauliflower rice
2 cups coconut milk
½ cup diced bittersweet chocolate (at least 80% cacao)
3 teaspoons cinnamon
1 teaspoon cayenne pepper
6 drops liquid stevia
¼ teaspoon sea salt
1 cup toasted walnuts

In a large saucepan, heat the cauliflower rice and half the coconut milk over medium heat.

Let simmer, stirring frequently, for about 5 to 10 minutes, until the "rice" is tender.

Add the remaining ingredients and stir until the chocolate is melted and well incorporated.

Let simmer for an additional 10 minutes and serve warm, or chill for a thicker, cold rice pudding.

Pears Poached in Red Wine with Vanilla Coconut Cream

Pears contain some powerful longevity-supporting compounds and lots of resistant starch, but, more important, they're delicious, especially when paired with citrus zest, star anise, and red wine. This recipe is a great way to turn even a slightly underripe pear into a sweet, rich, satisfying dessert, made more decadent by vanilla bean–infused coconut cream.

Serves 4

FOR THE PEARS:

2 cups red wine
2 cups coconut milk
⅔ cup erythritol powder
1 star anise
1 stick cinnamon
Peel of 1 orange, cut into large strips

(continued)

2 whole cloves

2 semiripe pears, peeled

FOR THE COCONUT CREAM:

1 can coconut cream, refrigerated for 24 hours

1 vanilla bean

Place a bowl and whisk in your refrigerator.

MAKE THE PEARS: In a large pot, heat the wine, coconut milk, erythritol powder, anise, cinnamon, orange peel, and cloves, stirring frequently until the erythritol powder dissolves and the mixture is simmering.

Add the pears and cook, covered, over medium-low heat until tender.

With a slotted spoon, remove the pears, spices, and orange peel from the poaching liquid. Set the pears aside and discard the spices and orange peel.

Turn heat to medium high and cook the poaching liquid until syrupy, then remove from heat.

MAKE THE COCONUT CREAM: Right before serving, add the chilled coconut cream to the refrigerated bowl. Scrape the vanilla bean and add the seeds to the coconut cream. Whip until stiff peaks have formed.

Serve half a pear per person, topped with the whipped coconut cream and a drizzle of the reduced poaching liquid.

Beverages

Walnut and Nutmeg "Horchata"

When I traveled in Spain, I loved the creamy, sweet drink called horchata, sold everywhere, so I set out to make my own—lectin free, of course. I think you'll agree that this slightly toasty, rich drink is every bit as good as the classic—and better for you, too.

Makes 4 cups

4 tablespoons millet
⅔ cup walnuts
¼ teaspoon cinnamon
½ teaspoon nutmeg
Zest of ½ orange
¼ cup erythritol powder
1 teaspoon vanilla extract

In a dry pan, toast the millet and walnuts over medium heat, stirring frequently.

(continued)

When the walnuts smell "toasty," remove from heat and let cool to room temperature.

Transfer the mixture to a spice grinder and pulse until powdery.

In a blender, combine the millet mixture, cinnamon, nutmeg, and orange zest. Pulse until a powder is formed.

Add 2 cups hot water and the erythritol powder and blend until dissolved, then add 2 cups cold water. Let sit at room temperature for 20 minutes, then chill for 4 hours.

Strain, add the vanilla extract, stir, and serve.

Green Smoothie

This is the perfect breakfast during your five-day "fast" and on free days as well. Add a little more water if the smoothie is too thick. You can make a triple batch and refrigerate for up to three days in a covered glass container.

Serves 1

1 cup chopped romaine lettuce
½ cup baby spinach
1 to 3 fresh mint sprigs, with stems
½ avocado
4 tablespoons freshly squeezed lemon juice
3 to 6 drops liquid stevia, to taste
¼ cup ice cubes
1 cup tap or filtered water

Place all the ingredients in a high-powered blender and blend on high until smooth and fluffy, adding more ice cubes if desired.

Condiments, Sauces, and Bases

Parmesan "Bone" Broth

This is an almost meaty-tasting broth that's a great way to use up Parmesan rinds, which are rich in longevity-boosting spermidine. It's great as a base for soups, in cauliflower rice risotto, or in any recipe calling for chicken broth.

Makes about 2 quarts

¼ cup extra-virgin olive oil
1 head garlic, peel on, cut in half across the middle
1 onion, cut into eighths
1 bunch of fresh thyme (about ½ cup)
1 small bunch of fresh parsley (about ¼ cup)
1 bay leaf
1 tablespoon whole black peppercorns
Zest of 1 lemon

1 cup dry white wine*

1 pound Parmesan cheese rinds**

In a large soup pot, heat the olive oil over medium-high heat.

Add the garlic (cut side down) and onion and cook until golden brown and fragrant.

Add the thyme and parsley and cook an additional 2 minutes.

Add the bay leaf, peppercorns, lemon zest, wine, and Parmesan rinds and cook, stirring frequently, until the rinds begin to soften and melt.

Add 9 cups water, reduce heat to low, cover, and simmer for 90 minutes.

Uncover and cook for an additional 20 minutes.

Strain and use immediately, refrigerate for up to 1 week, or freeze for up to 3 months.

*If you'd prefer not to use wine, use the juice of 1 lemon to add acidity and brightness.

**These freeze well, so save them up over time when you use fresh Parmesan.

Mushroom Miso Broth

This rich, Japanese-inspired broth is a great vegan option for folks who enjoy strong, meaty, umami-rich flavors. It's super good with Miracle Noodle brand pasta or as a broth to make a delicious, warming soup.

Makes about 2 quarts

¼ cup avocado oil
4 shallots, roughly chopped
1 head garlic, sliced in half along the "equator"
10 large shiitake mushrooms, thinly sliced
2 strips kombu (dried kelp)
¼ cup coconut aminos
¼ cup red or white miso
1 cup dry white wine
2 tablespoons yacon syrup

In a large soup pot, heat the oil over medium-high heat.

Add the shallots, garlic, and mushrooms and cook, stirring frequently, until the vegetables are tender.

Reduce heat to low. Add the kombu, coconut aminos, miso, wine, and yacon syrup and cook, stirring frequently, until the miso dissolves into the wine and the mixture is fragrant.

Add 8 cups water, cover, and simmer for 30 to 40 minutes.

Strain and use immediately, refrigerate for up to 2 weeks, or freeze for up to 3 months.

Miso Balsamic "Barbecue" Sauce

Love barbecue sauce but hate the lectins and sugar in conventional brands? This glaze is for you. It has a great balance of sweet, tangy, and savory flavors and is really incredible drizzled over grass-fed meats or roasted veggies.

Makes 2 cups

1 tablespoon avocado oil
2 large shallots, thinly sliced
½ teaspoon freshly ground black pepper
½ teaspoon cumin
¼ cup balsamic vinegar
¼ cup red miso paste
¼ cup yacon syrup
¼ cup coconut aminos
1 cup Parmesan "Bone" Broth (page 430)
½ cup apple cider vinegar

(continued)

In a saucepan, heat the oil over medium-high heat.

Add the shallots, pepper, and cumin and cook, stirring frequently, until the shallots are soft and caramelized.

Reduce heat to low and add the vinegar, miso paste, yacon syrup, and coconut aminos. Stir until the mixture thickens.

Add the broth and vinegar and cook, stirring frequently, until slightly thickened.

Strain into a jar and let cool before storing in the refrigerator for up to 1 week or in the freezer for up to 1 month.

To use, brush on wild-caught salmon or pasture-raised meat before cooking. You can also use this sauce as a marinade for veggies.

Miso Sesame Dressing

Inspired by the addictive salad dressing at my local sushi restaurant, this dressing is a little creamy and unbelievably flavorful. It's good on greens, drizzled over an egg or avocado, or even with roast veggies. It also works nicely as a fish marinade.

Makes 1 cup

½ cup white miso paste
⅓ cup plus 1 tablespoon water
¼ cup yacon syrup
3 tablespoons rice wine vinegar
2 teaspoons coconut aminos
3 tablespoons sesame oil
¼ cup minced scallions
1 clove garlic, minced

In a bowl, whisk together the miso paste and water until smooth.

Add the yacon syrup, vinegar, and coconut aminos and continue to whisk until smooth and well blended.

Drizzle in the sesame oil, whisking as you go to emulsify.

Fold in the scallions and garlic, then use on your

(continued)

favorite salad or store in the refrigerator for up to 2 weeks.

Let come to room temperature and shake well before using.

Basil Lentil "Pâté"

Though it doesn't taste like meaty pâté, this hearty spread or dip is super flavorful thanks to the mixture of savory miso and fresh basil. It's served here with Belgian endive leaves for dipping but is also great with raw broccoli florets or asparagus spears for extra fiber.

Makes 2 cups

2 tablespoons extra-virgin olive oil plus extra for drizzling
1 yellow onion, finely minced
1 tablespoon fresh thyme
1 teaspoon freshly ground black pepper
1 cup toasted walnuts
3 tablespoons red miso paste
1½ cups fresh basil leaves
¼ cup shredded Parmesan cheese or nutritional yeast

2 whole cloves garlic

1½ cups pressure-cooked lentils (green lentils preferred)

1 tablespoon coconut aminos

Water or chilled Parmesan "Bone" Broth (page 430), as needed

Belgian endive leaves, for serving

In a small skillet, heat the oil over medium-high heat.

Add the onion, thyme, and pepper and cook, stirring frequently, until the onions are translucent and tender. Set aside to let cool.

While the onion mixture is cooling, pulse the walnuts, miso paste, and basil leaves in a food processor fitted with an S blade.

Add the Parmesan, garlic, lentils, and coconut aminos and blend until a smooth paste is formed.

If needed, drizzle in water or broth a teaspoon at a time, until the mixture is smooth and silky.

Serve with Belgian endive leaves.

Afterword

The paradox of longevity comes down to this: No one's getting out alive. But you can die young at a ripe old age by assembling the right team. Or, perhaps more accurately, by assembling a village of trillions of inhabitants who want only one thing, and that is to preserve their beautiful home. Your team should also include a devoted collection of family members, friends, and animals who can provide you with emotional and social support and inspire you to stay active throughout your years.

But let's not forget the importance of your attitude on your quality of life. One of the traits I've observed in many of the "super old" people I've had the pleasure of knowing is an outlook I like to call pessimistic optimism. It's exemplified by the life-enhancing ability to shrug your shoulders at the inevitable bad things that happen and celebrate the many good things.

As an example, I'll leave you to contemplate Ruby, who is turning 102 as I write this. I've known Ruby for at least ten years. When I met her, this diminutive woman had hands and feet that were so gnarled and deformed by rheumatoid arthritis that my first question to her was, didn't it hurt horribly to walk? Her response was what I have come to expect from my great teachers: "Of course it hurts, but I can't do anything about it, so why pay it any attention?" She shrugged. And smiled. Simultaneously. And in her eyes I detected the intense sparkle of life that belied her true age. In her nineties, she taught chair yoga! She had a close network of friends. I began to look forward to every visit with Ruby, not just because she was such a joy to be around but also because I always learned something from her.

Year after year I would suggest that we try a few dietary changes to help with her rheumatoid arthritis, but she just wasn't interested. Shortly after her 100th birthday she found a lump in her breast that was cancerous and had to be removed. When we met after her surgery, I asked her what her plans were now that she was 100 and had cancer. Again, her answer didn't surprise me. She wanted to live to see her great-grandchildren graduate from high school and wouldn't let a little thing like cancer stand in her way. I asked if perhaps now was the time to try the dietary changes I had been suggesting, and she finally said, "Sure, Doc, let's give it a try."

When Ruby walked in the door for her 101st birthday visit, I was immediately struck by the appearance of her hands. The horribly deformed knuckles were now dramatically smaller, and her fingers and toes were noticeably straighter. A quick glance at her new blood work results showed me that her biomarkers for active rheumatoid arthritis, RF and Anti-CCP3, which were usually very elevated, were now normal. As in negative. I excitedly told her that her eating efforts had paid off, showing her the blood results and pointing to her relaxed hands. She said, "Yes, I've noticed my hands, but I have a bone to pick with you." With that she turned her hands and fingers toward the floor, and six rings fell off her fingers, clanging onto the tile. "I've got to get my rings resized!" The perfect pessimistic optimist in action.

Ruby's deaging is happening right before my eyes; now, at 102, she just keeps getting younger. And this is my hope for you. Through a combination of nutrition, lifestyle choices, a supportive community, and a mindset that seeks the positive while accepting the negative with humor and humility, we can all enjoy a full, vibrant life for as many years as we live on this earth.

Of course, Ruby's time to move onward will eventually come, just as my time and yours will. But until it does, I personally plan to nurture that vital spark that keeps her birthday candles lit year after year. Or as Jason Mraz would say, "May the best of your todays be the worst of your tomorrows." Die young, dear reader, at a ripe old age!

Acknowledgments

Just as it takes a village to live well to a ripe old age, it took a village of talented individuals to help get this book into your hands. My collaborator, Jodi Lipper, was able to tame and make clearer my very lengthy prose and writing style. Recipe developer Kathryn Holzhauer has once again provided us with incredible and inventive lectin-free recipes, and Irina Skoeries provided the five-day vegan fast-mimicking diet.

The team at HarperWave did it again without so much as a hiccup in getting this project completed in record time. Thanks to my now longtime publisher, Karen Rinaldi; to Brian Perrin, director of marketing; and to Yelena Nesbit, director of publicity. I'm also grateful to Milan Bozic for the beautiful cover, and to Haley Swanson and Nikki Baldauf for handling the production details flawlessly. And of course, thank you to my dear editor extraordinaire, Julie Will, who lovingly beat me and *The Plant Paradox* into the major bestseller that has changed so many lives for the better, and whose editorial eye (and unabashed honesty) I have relied on for every book since.

Thanks, too, to my longtime agent and early believer, Shannon Marven, president of Dupree Miller; my attorney and longtime friend and supporter, Dave Baron; his associate, Ini Ghidirmic; and my accountant, Joyce Ohmura, who were able to corral all these disparate entities into a beautiful finished product.

As I said in *The Plant Paradox*, I cannot thank individually the entire five hundred-plus people at GundryMD who have made me and GundryMD.com the trusted source for health and supplement advice for millions of people daily, but I have to single out Lanee Lee Neil, who for the past year has daily—weekends as well—lorded over me and my brand. Lanee, I couldn't have done it without you! Likewise, Lauren Newhouse and her team of publicists, including Rebecca Reinbold and Jessica Hofmann at Stanton Company, keep me and GundryMD in the spotlight daily. Thank you, all.

And speaking of couldn't have done it without you, heartfelt thanks to my entire staff at the International Heart and Lung Institute and the Centers for Restorative Medicine in Palm Springs and Santa Barbara, California! As if things weren't busy enough before *The Plant Paradox*, wow, did you guys step up to the plate! Directed by Susan Lokken, my loyal team of Adda Harris; Tanya Marta; Cindy Crosby (who singlehandedly keeps the office afloat monetarily); Donna Fitzgerald; my daughter, Melissa Perko; of course, the "Blood Suckers," led by Laurie Acuna and Lynn Visk; and my physician assistant, the fantastic Mitsu Killion-Jacobo. How you keep the chaos controlled is a true testament to your loyalty to the cause of making our patients well.

Speaking of controlling chaos, my real rock in all of this is my wife, Penny, who, along with our three dogs, never lets me forget that, when the sun rises each day, I'm really only a dog walker and if I do that job well, the rest of the day will follow suit!

Finally, none of this would be possible without my patients and you, my readers. Thank you for your trust in me and my team as we together try to maximize our collective knowledge and health.

Notes

Introduction: This Is a Test

1. Katherine Harmon, "Cracked Corn: Scientists Solve Maize's Genetic Maze," *Scientific American*, November 19, 2009, https://www.scientificamerican.com/article/corn-genome-cracked/.
2. "The Most Genes in an Animal? Tiny Crustacean Holds the Record," National Science Foundation, February 3, 2011, https://www.nsf.gov/news/news_summ.jsp?cntn_id=118530.
3. Ann Gibbons, "Microbes in Our Guts Have Been with Us for Millions of Years," *Science*, July 21, 2016, http://www.sciencemag.org/news/2016/07/microbes-our-guts-have-been-us-millions-years.
4. Bruce Goldman, "Low-Fiber Diet May Cause Irreversible Depletion of Gut Bacteria over Generations," Stanford

Medicine News Center, January 13, 2016, https://med.stanford.edu/news/all-news/2016/01/low-fiber-diet-may-cause-irreversible-depletion-of-gut-bacteria.html.

5. "NIH Human Microbiome Project Defines Normal Bacteria Makeup of the Body," National Institutes of Health, June 13, 2012, https://www.nih.gov/news-events/news-releases/nih-human-microbiome-project-defines-normal-bacterial-makeup-body.
6. Emanuela Viggiano, Maria Pina Mollica, Lillà Lionetti, et al., "Effects of an High-Fat Diet Enriched in Lard or in Fish Oil on the Hypothalamic Amp-Activated Protein Kinase and Inflammatory Mediators," *Frontiers in Cellular Neuroscience* 10 (June 2016): 150, https://www.ncbi.nlm.nih.gov/pmc/articles/PMC4899473/.
7. Honor Whiteman, "CDC: Life Expectancy in the US Reaches Record High," Medical News Today, October 8, 2014, http: http://www.medicalnewstoday.com/articles/283625.php.
8. Mike Stobbe, "U.S. Life Expectancy Will Likely Decline for Third Straight Year," Bloomberg, May 23, 2018, https://www.bloomberg.com/news/articles/2018-05-23/with-death-rate-up-us-life-expectancy-is-likely-down-again.
9. Duke Health, "Physical Declines Begin Earlier Than Expected among U.S. Adults," ScienceDaily, July 21, 2016, https://www.sciencedaily.com/releases/2016/07/160721144805.htm.

10. Jean Epstein and Steven Gundry, "OP-055 Twelve Year Followup for Manging Coronary Artery Disease Using a Nutrigenomics Based Diet and Supplement Program and Quarterly Assessment of Biomarkers," *The American Journal of Cardiology* 115, no. 1 (March 2015): https://www.researchgate.net/publication/273910214_OP-055_Twelve_Year_Followup_for_Managing_Coronary_Artery_Disease_Using_a_Nutrigenomics_Based_Diet_and_Supplement_Program_and_Quarterly_Assessment_of_Biomarkers.
11. Steven Gundry, "Modifying the Gut Microbiome with Polphenols and a Lectin Limited Diet Improves Endothelial Function," *Atherosclerosis* 252, no. 167 (September 2016): https://www.researchgate.net/publication/308575652_Modifying_the_gut_microbiome_with_polyphenols_and_a_lectin_limited_diet_improves_endothelial_function.

Chapter 1: Ancient Genes Control Your Fate

1. Dariush Mozaffarian, Tao Hao, Eric B. Rimm, et al., "Changes in Diet and Lifestyle and Long-Term Weight Gain in Women and Men," *New England Journal of Medicine* 364, no. 25 (June 2011): 2392–404, https://www.nejm.org/doi/full/10.1056/NEJMoa1014296?query OC&smid ytcore-ios-share#t=articleTop.
2. Daphna Rothschild, Omer Weissbrod, Elad Barkan, et al.,

"Environment Dominates over Host Genetics in Shaping Human Gut Microbiota," *Nature* 555 (March 2018): 210–15, https://www.nature.com/articles/nature25973.

3. Gaorui Bian, Gregory B. Gloor, Aihua Gong, et al., "The Gut Microbiota of Healthy Aged Chinese Is Similar to That of the Healthy Young," *mSphere* 2, no. 5 (September–October 2017): https://msphere.asm.org/content/msph/2/5/e00327-17.full.pdf.
4. Elena Biagi, Claudio Franceschi, Simone Rampelli, et al., "Gut Microbiota and Extreme Longevity," *Current Biology: CB* 26, no. 11 (May 2016): https://www.researchgate.net/publication/303027047_Gut_Microbiota_and_Extreme_Longevity.
5. Adolfo Sanchez-Blanco, Alberto Rodríguez-Matellán, Ana González-Paramás, et al., "Dietary and Microbiome Factors Determine Longevity in *Caenorhabditis elegans*," *Aging* 8, no. 7 (July 2016): 1513–30, https://www.ncbi.nlm.nih.gov/pmc/articles/PMC4993345/.
6. University of Alabama at Birmingham, "Scientists Reverse Aging-Associated Skin Wrinkles and Hair Loss in a Mouse Model: A Gene Mutation Causes Wrinkled Skin and Hair Loss; Turning Off That Mutation Restores the Mouse to Normal Appearance," ScienceDaily, July 20, 2018, https://www.sciencedaily.com/releases/2018/07/180720112808.htm.
7. Jan Gruber and Brian K. Kennedy, "Microbiome and Longevity: Gut Microbes Send Signals to Host Mitochon-

dra," *Cell* 169, no. 7 (June 2017): 1168–69, http://www.cell.com/cell/fulltext/S0092-8674(17)30641-4.

8. Bing Han, Pryia Sivaramakrishnan, Chih-Chun J. Lin, et al., "Microbial Genetic Composition Tunes Host Longevity," *Cell* 169, no. 7 (June 2017): 1249–62, https://www.sciencedirect.com/science/article/pii/S009286741730627X.

9. Elaine Patterson, John F. Cryan, Gerald F. Fitzgerald, et al., "Gut Microbiota, the Pharmabiotics They Produce and Host Health," *Proceedings of the Nutrition Society* 73, no. 4 (November 2014): 477–89, https://www.cambridge.org/core/journals/proceedings-of-the-nutrition-society/article/gut-microbiota-the-pharmabiotics-they-produce-and-host-health/4961D7293641D4FC3255468A22C7FF66.

10. Jin Li, Shaoqiang Lin, Paul M. Vanhoutte, et al., "*Akkermansia Muciniphila* Protects Against Atherosclerosis by Preventing Metabolic Endotoxemia-Induced Inflammation in *Apoe* Mice," *Circulation* 133, no. 24 (April 2016): 2434–46, https://www.ahajournals.org/doi/full/10.1161/circulationaha.115.019645.

11. Ming Fu, Weihua Zhang, Lingyun Wu, et al., "Hydrogen Sulfide (H_2S) Metabolism in Mitochondria and Its Regulatory Role in Energy Production," *Proceedings of the National Academy of Sciences of the United States of America* 109, no. 8 (February 2012): 2943–48, http://www.pnas.org/content/109/8/2943.

12. Tewodros Debebe, Elena Biagi, Matteo Soverini, et al., "Unraveling the Gut Microbiome of the Long-Lived Naked Mole-Rat," *Scientific Reports* 7 (August 2017): https://www.nature.com/articles/s41598-017-10287-0.

13. M. Bernstein, "Bifidobacteria lpngum, Roseburia, F. prausnitzii (and Akkermansia) Made Us Human (None of These Eat Raw Potato Starch) (Part 1) NSFW," Animal Pharm, November 9, 2014, http://drbganimalpharm.blogspot.com/2014/11/bifidobacteria-longum-roseburia-f.html.

14. Anna Azvolinsky, "Primates, Gut Microbes Evolved Together," *The Scientist*, July 21, 2016, http://mobile.the-scientist.com/article/46603/primates-gut-microbes-evolved-together.

15. Yolanda Sanz, "Effects of a Gluten-Free Diet on Gut Microbiota and Immune Function in Healthy Adult Humans," *Gut Microbes* 1, no. 3 (March 2010): 135–37, http://www.ncbi.nlm.nih.gov/pmc/articles/PMC3023594/.

16. David R. Montgomery and Anne Biklé, *The Hidden Half of Nature: The Microbial Roots of Life and Health* (New York: W. W. Norton, 2016), cited a study by Lee S. Gross, Li Li, Earl S. Ford, and Simin Liu, "Increased Consumption of Refined Carbohydrates and the Epidemic of Type 2 Diabetes in the United States: An Ecologic Assessment," *American Journal of Clinical Nutrition* 79 (2004): 774–79, https://cwru.pure.elsevier.com/en/publications/increased-consumption-of-refined-carbohydrates-and-the-epidemic-o-3.

17. University of Colorado at Boulder, "Personal Care Products Contribute to a Pollution 'Rush Hour': Emissions from Products Such as Shampoo and Perfume Are Comparable to the Emissions from Auto Exhaust," ScienceDaily, April 30, 2018, https://www.sciencedaily.com/releases/2018/04/180430131828.htm.

18. Roddy Scheer and Doug Moss, "Dirt Poor: Have Fruits and Vegetables Become Less Nutritious?," *Scientific American* (no date), http://www.scientificamerican.com/article/soil-depletion-and-nutrition-loss/.

19. Martin J. Blaser, *Missing Microbes: How the Overuse of Antibiotics Is Fueling Our Modern Plagues* (New York: Henry Holt, 2014).

20. Peter J. Turnbaugh, Micah Hamady, Tanya Yatsunenko, et al., "A Core Gut Microbiome in Obese and Lean Twins," *Nature* 457, no. 7228 (January 22, 2009): 480–84, https://www.nature.com/articles/nature07540.

21. John J. Gildea, David A. Roberts, and Zachary Bush, "Protective Effects of Lignite Extract Supplement on Intestinal Barrier Functions in Glyphosate-Mediated Tight Junction Injury," *Journal of Clinical Nutrition and Diabetics* 3, no. 1 (January 2017): 1–6, http://clinical-nutrition.imedpub.com/abstract/protective-effects-of-lignite-extract-supplement-on-intestinal-barrier-function-in-glyphosatemediated-tight-junction-injury-18161.html.

22. Shannon Van Hoesen, "World Health Organization Labels Glyphosate Probable Carcinogen," EWG, March 2015,

http://www.ewg.org/release/world-health-organization-labels-glyphosate-probable-carcinogen.

23. S. Parvez, R. R. Gerona, C. Proctor, et al., "Glyphosate Exposure in Pregnancy and Shortened Gestational Length: A Prospective Indiana Birth Cohort Study," *Environmental Health* 17, no. 1 (2018): 23, https://ehjournal.biomedcentral.com/articles/10.1186/s12940-018-0367-0.

24. A. Gore, V. Chappell, S. Fenton, et al., "EDC-2: The Endocrine Society's Second Scientific Statement on Endocrine-Disrupting Chemicals," *Endocrine Reviews* 36, no. 6, (December 2015): E1–E150, http://www.ncbi.nlm.nih.gov/pubmed/26544531.

25. Pheruza Tarapore, Jun Ying, Bin Ouyang, et al., *PLOS One* 9, no. 3 (March 2014): 1–11, http://journals.plos.org/plosone/article?id=10.1371/journal.pone.0090332.

26. Jan Gruber and Brian K. Kennedy, "Microbiome and Longevity: Gut Microbes Send Signals to Host Mitochondria," *Cell* 169, no. 7 (June 2017): 1168–69, http://www.cell.com/cell/fulltext/S0092-8674(17)30641-4.

27. Yaw A. Nyame, Adam B. Murphy, Diana K. Bowen, et al., "Assocations Between Serum Vitamin D and Adverse Pathology in Men Undergoing Radical Prostatectomy," *Journal of Clinical Oncology* 34, no. 12 (April 2016): 1345–50, http://ascopubs.org/doi/10.1200/JCO.2015.65.1463.

28. M. B. Abou-Donia, E. M. El-Masry, A. A. Abdel-Rahman, et al., "Splenda Alters Gut Microflora and Increases Intestinal P-Glycoprotein and Cytochrome P-450 in Male Rats," *Journal of Toxicology Environmental Health* 71, no. 21 (2008): 1415–29, https://www.ncbi.nlm.nih.gov/m/pubmed/18800291/.

Chapter 2: Protect and Defend

1. Claudio Franceschi and Judith Campisi. "Chronic Inflammation (Inflammaging) and Its Potential Contribution to Age-Associated Diseases," Journals of Gerontology: Series A 69, no. 1 (June 2014): S4–S5, https://www.ncbi.nlm.nih.gov/m/pubmed/24833586/.
2. University of Pittsburgh Medical Center, "Difference Between Small and Large Intestine," Children's Hospital of Pittsburgh Educational Resources, 2018, http://www.chp.edu/our-services/transplant/intestine/education/about-small-large-intestines.
3. Francie Diep, "Human Gut Has the Surface Area of a Studio Apartment," *Popular Science*, April 23, 2014, http://www.popsci.com/article/science/human-gut-has-surface-area-studio-apartment.
4. "Digestive 6," Quizlet, 2018, https://quizlet.com/11845442/digestive-6-flash-cards/.
5. Franceschi, and Campisi, "Chronic Inflammation (Inflammaging) and Its Potential Contribution to Age-Associated Diseases."
6. S. Manfredo Vieira, Michael Hiltensperger, V. Kumar, et al., "Translocation of a Gut Pathobiont Drives Autoimmunity in Mice and Humans," *Science* 359, no. 6380

(March 9, 2018): 1156–61, http://science.sciencemag.org/content/359/6380/1156.

7. Steven R. Gundry, "Abstract P238: Remission/Cure of Autoimmune Diseases by a Lectin Limite Diet Supplemented with Probiotics, Prebiotics, and Polyphenols," *Circulation* 137, no. 1, (June 2018): 238, https://www.ahajournals.org/doi/abs/10.1161/circ.137.suppl_1.p238.
8. Jawahar L. Mehta, Tom G. P. Saldeen, and Kenneth Rand, "Interactive Role of Infection, Inflammation and Traditional Risk Factors in Atherosclerosis and Coronary Artery Disease," *Journal of the American College of Cardiology* 31, no. 6 (May 1998): 1217–25, https://www.sciencedirect.com/science/article/pii/S073510979800093X.
9. Robert J. F. Laheij, Miriam C. J. M. Sturkenboom, Robert-Jan Hassing, et al., "Risk of Community-Acquired Pneumonia and Use of Gastric Acid–Suppressive Drugs," *Journal of the American Medical Association* 292, no. 16 (October 2004): 1955–60, http://jama.jamanetwork.com/article.aspx?articleid=199672.
10. Jan Bures, Jiri Cyrany, Darina Kohoutová, et al., "Small Intestinal Bacterial Overgrowth Syndrome," *World Journal of Gastroenterology* 16, no. 24 (June 2010): 2978–90, https://www.researchgate.net/publication/44696633_Small_intestinal_bacterial_overgrowth_syndrome.
11. Medical College of Georgia at Augusta University, "Drinking Baking Soda Could Be an Inexpensive, Safe Way to Combat Autoimmune Disease," ScienceDaily,

April 25, 2018, https://www.sciencedaily.com/releases/2018/04/180425093745.htm.

12. S. Y. Tai, C. Chien, D. Wu, et al., "Risk of Dementia from Proton Pump Inhibitor Use in Asian Population: A Nationwide Cohort Study in Taiwan," *PLOS One* 12, no. 2 (February 2017): https://www.ncbi.nlm.nih.gov/pubmed/28199356.

13. Dennis Thompson, "Popular Heartburn Drugs Linked to Risk of Dementia," *USA Today*, February 16, 2016, http://www.usatoday.com/story/news/health/2016/02/16/popular-heartburn-drugs-linked-to-risk-of-dementia/80442834/.

14. Benjamin Lazarus, Yuan Chen, Francis P. Wilson, et al., "Proton Pump Inhibitor Use and Risk of Chronic Kidney Disease," *Journal of the American Medical Association* 176, no. 2 (February 2016): 238–46, https://www.ncbi.nlm.nih.gov/pmc/articles/PMC4772730/.

15. U. Engel, D. Breborowicz, T. Bøg-Hansen, et al., "Lectin Staining of Renal Tubules in Normal Kidney," *Applied Mathematics and Information Sciences* 105, no. 1 (January 1997): 31–34, http://www.ncbi.nlm.nih.gov/m/pubmed/9063498/.

16. George Institute for Global Health, "Diabetes Raises Risk of Cancer, with Women at Even Greater Likelihood, a Major New Study Has Found," ScienceDaily, July 19, 2018, https://www.sciencedaily.com/releases/2018/07/180719195650.htm.

17. "Statistics About Diabetes," American Diabetes Association, http://www.diabetes.org/diabetes-basics/statistics/.

18. IOS Press, "Insulin Resistance Under-Diagnosed in Non-Diabetics with Parkinson's Disease," ScienceDaily, August 2, 2018, https://www.sciencedaily.com/releases/2018/08/180802151525.htm.

19. Edward J. Calabrese and Linda A. Baldwin, "Radiation Hormesis: Origins, History, Scientific Foundations," *BELLE Newsletter* 8, no. 2 (December 1999): http://dose-response.org/wp-content/uploads/2014/05/www_belleonline_com_newsletters_volume8_vol8_2_html.pdf.

20. K. J. Mukamal, S. E. Chiuve, and E. B. Rimm, "Alcohol Consumption and Risk for Coronary Heart Disease in Men with Healthy Lifestyles," *Archives of Internal Medicine* 166, no. 19 (October 2006): 2145–50, https://www.ncbi.nlm.nih.gov/pubmed/17060546.

21. Centre National de la Recherche Scientifique, "Eating Less Enables Lemurs to Live Longer," ScienceDaily, April 5, 2018, https://www.sciencedaily.com/releases/2018/04/180405093241.htm.

22. Sara Gelino, Jessica T. Chang, Caroline Kumsta, et al., "Intestinal Autophagy Improves Healthspan and Longevity in *C. elegans* during Dietary Restriction," *PLOS Genetics* 12, no. 8 (July 2016): http://journals.plos.org/plosgenetics/article?id=10.1371/journal.pgen.1006135.

23. Louise E. Tailford, Emmanuelle H. Crost, Devon Kavanaugh, et al., "Mucin Glycan Foraging in the

Human Gut Microbiome," *Frontiers in Genetics* 6, no. 8 (March 2015): https://www.frontiersin.org/articles/10.3389/fgene.2015.00081/full.

24. Jin Li, Shaoqiang Lin, Paul M. Vanhoutte, et al., "*Akkermansia Muciniphila* Protects Against Atherosclerosis by Preventing Metabolic Endotoxemia-Induced Inflamation in *Apoe* Mice," *Circulation* 133, no. 24 (June 2016): 2434–46, https://www.ncbi.nlm.nih.gov/m/pubmed/27143680/?i=4&from=/23671105/related.

25. Moran Yassour, Mi Young Lim, Hyun Sun Yun, et al., "Sub-Clinical Detection of Gut Microbial Biomakers of Obesity and Type 2 Diabetes," *Genome Medicine* 8, no. 16 (February 2016): https://genomemedicine.biomedcentral.com/articles/10.1186/s13073-016-0271-6.

26. Scott C. Anderson, "Can Mucus-Loving Bacteria Cure Obesity and Diabetes?," Notch by Notch, May 16, 2013, http://notchbynotch.com/can-mucus-loving-bacteria-cure-obesity-and-diabetes/.

27. J. de la Cuesta-Zuluaga, N. T. Mueller, V. Corrales-Agudelo, et al., "Metformin Is Associated with Higher Relative Abundance of Mucin-Degrading Akkermansia Muciniphila and Several Short-Chain Fatty Acid-Producing Microbiota in the Gut," *Diabetes Care* 40, no. 1 (January 2017): https://www.ncbi.nlm.nih.gov/pubmed/27999002.

28. Ibid.

29. M. Carmen Collado, Muriel Derrien, Erika Isolauri, et al., "Intestinal Integrity and *Akkermansia muciniphila*, a Mucin-Degrading Member of the Intestinal Microbiota

Present in Infants, Adults, and the Elderly," *Applied and Environmental Microbiology* 73, no. 23 (2007): 7767–70, https://aem.asm.org/content/73/23/7767.

30. X. Gao, Q. Xie, P. Kong, et al., "Polyphenol- and Caffeine-Rich Postfermented Pu-erh Tea Improves Diet-Induced Metabolic Syndrome by Remodeling Intestinal Homeostasis in Mice," *Infection and Immunity* 86, no. 1 (January 2018): https://www.ncbi.nlm.nih.gov/m/pubmed/29061705/?i=4&from=/25080446/related.

31. Laura García-Prat, Marta Martínez-Vicente, Eusebio Perdiguero, et al., "Autophagy Maintains Stemness by Preventing Senescence," *Nature* 529, no. 1 (January 2016): 37–57, http://twin.sci-hub.tw/6695/3bb2aef883e8dde32b63cc52a0f897cc/garcaprat2016.pdf.

32. University of Southern California, "Fasting Triggers Stem Cell Regeneration of Damaged, Old Immune System," ScienceDaily, June 5, 2014, https://www.sciencedaily.com/releases/2014/06/140605141507.htm.

33. S. Melanie Lee, Gregory P. Donaldson, Zbigniew Mikulski, et al., "Bacterial Colonization Factors Control Specificity and Stability of the Gut Microbiota," *Nature* 501, no. 7467 (September 19, 2013): 426–29, https://www.researchgate.net/profile/Klaus_Ley/publication/255975522_Bacterial_colonization_factors_control_specificity_and_stability_of_the_gut_microbiota/links/0f31753cd1efbd6ade000000/Bacterial-colonization-factors-control-specificity-and-stability-of-the-gut-microbiota.pdf.

34. Eugene Kang, Mitra Yousefi, and Samantha Gruenheid, "R-Spondins Are Expressed by the Intestinal Stroma and Are Differently Regulated during *Citrobactoer rodentium*- and DSS-Induced Colitis in Mice," *PLOS One* 11, no. 4, (April 2016): http://journals.plos.org/plosone/article?id=10.1371/journal.pone.0152859.

35. Karla A. Mark, Kathleen J. Dumas, Dipa Bhaumik, et al., "Vitamin D Promotes Protein Homeostasis and Longevity via the Stress Response Pathway Genes skn-1, ire-1, and xbp-1," *Cell Reports* 17, no. 5 (October 2016): 1227–37, https://www.researchgate.net/publication/309455488_Vitamin_D_Promotes_Protein_Homeostasis_and_Longevity_via_the_Stress_Response_Pathway_Genes_skn-1_ire-1_and_xbp-1.

36. Javeria Saleem, Rubeena Zakar, Muhammad Z. Zakar, et al., "High-Dose Vitamin D3 in the Treatment of Severe Acute Malnutrition: A Multicenter Double-Blind Randomized Controlled Trial," *American Journal of Clinical Nutrition* 107, no. 5 (May 2018): 725–33, https://academic.oup.com/ajcn/article-abstract/107/5/725/4990735?redirectedFrom=fulltext.

37. I. Flores, A. Canela, E. Vera, et al., "The Longest Telomeres: A General Signature of Adult Stem Cell Compartments," *Genes and Development* 22, no. 5 (March 2008): 654–67, https://www.ncbi.nlm.nih.gov/m/pubmed/18283121/.

38. J. Brent Richards, Ana M. Valdes, Jeffrey P. Gardner, et al., "Higher Serum Vitamin D Concentrations Are Associated with Longer Leukocyte Telomere Length in Women,"

American Journal of Clinical Nutrition 86, no. 5 (November 2007): 1420–25, https://www.ncbi.nlm.nih.gov/pmc/articles/PMC2196219/.

39. Cedric F. Garland, Christine B. French, Leo L. Baggerly, and Robert P. Heaney, "Vitamin D Supplement Doses and Serum 25-Hydroxyvitamin D in the Range Associated with Cancer Prevention," *Anticancer Research* 31, no. 2 (February 2011): 607–11.

40. Samuel A. Smits, Jeff Leach, Erica D. Sonnenburg, et al., "Seasonal Cycling in the Gut Microbiome of the Hadza Hunter-Gatherers of Tanzania," *Science* 357, no. 6353 (August 2017): 802–6, http://science.sciencemag.org/content/357/6353/802.

41. Columbia University's Mailman School of Public Health, "What the Gorilla Microbiome Tells Us about Evolution and Human Health: Researchers Find the Microbiomes of Wild Gorillas Shift Seasonally When Once a Year They Switch from Eating Fibrous Leaves to Eating Fruit," ScienceDaily, May 3, 2018, https://www.sciencedaily.com/releases/2018/05/180503085553.htm.

42. J. L. Broussard, D. A. Ehrmann, E. van Cauter, et al., "Impaired Insulin Signaling in Human Adipocytes after Experimental Sleep Restriction: A Randomized, Crossover Study," *Annals of Internal Medicine* 157, no. 8 (October 2012): 549–57, https://www.ncbi.nlm.nih.gov/m/pubmed/23070488/.

43. Oren Froy, "Circadian Rhythms, Aging, and Life Span in Mammals," *Physiology* 26, no. 1 (August 2011):

225–35, https://www.physiology.org/doi/abs/10.1152/physiol.00012.2011?url_ver=Z39.88-2003&rfr_id=ori%3Arid%3Acrossref.org&rfr_dat=cr_pub%3Dpubmed&.

44. Hung-Chun Chang and Leonard Guarente, "SIRT1 Mediates Central Circadian Control in the SCN by a Mechanism That Decays with Aging," *Cell* 153, no. 7 (2013): 1448–60, https://www.cell.com/abstract/S0092-8674%2813%2900594-1.

45. Maria Pina Mollica, Giuseppina Mattace Raso, Gina Cavaliere, et al., "Butyrate Regulates Liver Mitochondrial Function, Efficiency and Dynamics in Insulin-Resistant Obese Mice," *American Diabetes Association* 66 (May 2017): 1405–18, http://diabetes.diabetesjournals.org/content/66/5/1405.

46. Satya Sree N. Kolar, Rola Barhoumi, Joanne R. Lupton, et al., "Docosahexaenoic Acid and Butyrate Synergistically Induce Colonocyte Apoptosis by Enhancing Mitochondrial Ca^{2+} Accumulation," *American Association for Cancer Research* 67, no. 11 (June 2007): 5561–68, http://cancerres.aacrjournals.org/content/67/11/5561.long.

47. M. N. Ebert, G. Beyer-Sehlmeyer, U. M. Liegibel, et al., "Butyrate Induces Glutathione S-Transferase in Human Colon Cells and Protects from Genetic Damage by 4-Hydroxy-2-Nonenal," *Nutrition and Cancer* 41, nos. 1–2 (2001): 154–64, https://www.ncbi.nlm.nih.gov/m/pubmed/12094619/.

48. Megan W. Bourassa, Ishraq Alim, Scott J. Bultman, et al.,

"Butyrate, Neuroepigenetics and the Gut Microbiome: Can a Higher Fiber Diet Improve Brain Health?," *Neuroscience Letters* 625 (June 2016): 56–63, https://www.sciencedirect.com/science/article/pii/S0304394016300775.

49. N. Govindarajan, R. C. Agis-Balboa, J. Walter, et al., "Sodium Butyrate Improves Memory Function in an Alzheimer's Disease Mouse Model When Administered at an Advanced Stage of Disease Progression," *Journal of Alzheimer's Disease* 26, no. 1 (2011): 187–97, https://www.ncbi.nlm.nih.gov/m/pubmed/21593570/.

50. Huating Li, Zhanguo Gao, Jin Zhang, et al., "Sodium Butyrate Stimulates Expression of Fibroblast Growth Factor 21 in Liver by Inhibition of Histone Deacetylase 3," *Diabetes* 61 (April 2012): 797–806, http://diabetes.diabetesjournals.org/content/61/4/797.

51. Jean-Paul Buts, Nadine de Keyser, Jaroslaw Kolanowski, et al., "Maturation of Villus and Crypt Cell Functions in Rat Small Intestine. Role of Dietary Polyamines," *Digestive Diseases and Sciences* 38, no. 6 (1993): 1091–98, https://link.springer.com/article/10.1007/BF01295726.

52. Mitsuharu Matsumoto, Shin Kurihara, Ryoko Kibe, et al., "Longevity in Mice Is Promoted by Probiotic-Induced Suppression of Colonic Senescence Dependent on Upregulation of Gut Bacterial Polyamine Production," *PLOS One* 6, no. 8 (August 16, 2011): https://journals.plos.org/plosone/article?id=10.1371/journal.pone.0023652.

53. Fei Yue, Wenjiao Li, Jing Zou, et al., "Spermidine Prolongs Lifespan and Prevents Liver Fibrosis and Hepatocellular Carcinoma by Activating MAP1S-Mediated Autophagy," *American Association for Cancer Research* 77, no. 11 (April 2017): 1–32, http://cancerres.aacrjournals.org/content/early/2017/04/06/0008-5472.CAN-16-3462.

54. J. Mercola, "Why Aged Cheese and Mushrooms Are So Good for Your Heart (and Make You Live Longer Too)," The Science of Eating, 2016, http://thescienceofeating.com/2017/08/17/aged-cheese-mushrooms-good-heart-make-live-longer/.
55. Eugenia Morselli, Guillermo Mariño, Martin V. Bennetzen, et al., "Spermidine and Resveratrol Induce Autophagy by Distinct Pathways Converging on the Acetyloproteome," *Journal of Cell Biology* 192, no. 4 (February 2011): 615–29, jcb.rupress.org/content/192/4/615.full.

Chapter 3: What You Think Is Keeping You Young Is Probably Making You Old

1. M. A. Martínez-González, M. García-López, M. Bes-Rastrollo, et al., "Mediterranean Diet and the Incidence of Cardiovascular Disease: A Spanish Cohort," *Nutrition, Metabolism, and Cardiovascular Disease* 21, no. 4 (April 2011): 237–44, http://www.ncbi.nlm.nih.gov/pubmed/20096543.
2. M. Schünke, U. Schumacher, and B. Tillmann, "Lectin-Binding in Normal and Fibrillated Articular Cartilage of Human Patellae," *Virchows Archiv A, Pathological*

Anatomy and Histopathology 207, no. 2 (1985): 221–31, http://www.ncbi.nlm.nih.gov/m/pubmed/3927585/?i=5&from=/23214295/related.

3. Claudia Sardu, Eleonora Cocco, Alessandra Mereu, et al., "Population Based Study of 12 Autoimmune Diseases in Sardinia, Italy: Prevalence and Comorbidity," *PLOS One* 137, no. 10 (March 2012): http://journals.plos.org/plosone/article?id=10.1371/journal.pone.0032487.
4. Bradley J. Willcox, D. Craig Willcox, Hidemi Todoriki, et al., "Caloric Restriction, the Traditional Okinawan Diet, and Healthy Aging: The Diet of the World's Longest-Lived People and Its Potential Impact on Morbidity and Life Span," *New York Academy of Sciences* 1114 (2007): 434–55, http://www.okicent.org/docs/anyas_cr_diet_2007_1114_434s.pdf.
5. Caroline L. Bodinham, Gary S. Frost, M. Denise Robertson, "Acute Ingestion of Resistant Starch Reduces Food Intake in Healthy Adults," *British Journal of Nutrition* 103, no. 6 (March 2010): 917–22, http://journals.cambridge.org/action/displayAbstract?fromPage=online&aid=7358712&fileId=S0007114509992534.
6. Kevin C. Maki, Christine L. Pelkman, E. Terry Finocchiaro, et. al., "Resistant Starch from High-Amylose Maize Increases Insulin Sensitivity in Overweight and Obese Men," The Journal of Nutrition 142, no. 4 (April 1, 2012): 717–23.
7. A. C. Nilsson, E. M. Ostman, J. J. Holst, et al., "Including Indigestible Carbohydrates in the Evening Meal of

Healthy Subjects Improves Glucose Tolerance, Lowers Inflammatory Markers, and Increases Satiety after a Subsequent Standardized Breakfast," *Journal of Nutrition* 138, no. 4 (April 2008): 732–39, http://www.ncbi.nlm.nih.gov/pubmed/18356328.

8. Katri Korpela, Harry J. Flint, Alexandra M. Johnstone, et al., "Gut Microbiota Signatures Predict Host and Microbiota Responses to Dietary Interventions in Obese Individuals," *PLOS One* 10 (September 2014): http://www.oalib.com/references/8108647.

9. Patricia Lopez-Legarrea, Rocio de la Iglesia, Itziar Abete, et al., "The Protein Type within a Hypocaloric Diet Affects Obesity-Related Inflammation: The RESMENA Project," *Nutrition* 30, no. 4 (April 2014): 424–29, https://www.sciencedirect.com/science/article/pii/S0899900713004413.

10. Megan Durisin and Shruti Singh, "Americans Will Eat a Record Amount of Meat in 2018," Bloomberg, January 2, 2018, https://www.bloomberg.com/news/articles/2018-01-02/have-a-meaty-new-year-americans-will-eat-record-amount-in-2018.

11. Luigi Fontana, Edward P. Weiss, Dennis T. Villareal, et al., "Long-Term Effects of Calorie or Protein Restriction on Serum IGF-1 and IGFBP-3 Concentration in Humans," *Aging Cell* 7, no. 5 (October 2008): 681–87, http://www.ncbi.nlm.nih.gov/pmc/articles/PMC2673798/.

12. Crystal S. Conn and Shu-Bing Qian, "mTOR Signaling in Protein Homeostasis: Less Is More?," *Cell Cycle* 10, no. 12

(June 2015): 1940–47, http://www.ncbi.nlm.nih.gov/pmc/articles/PMC3154417/.

13. Michael J. Orlich, Pramil N. Singh, Joan Sabaté, et al., "Vegetarian Dietary Patterns and Mortality in Adventist Health Study 2," *JAMA Internal Medicine* 173, no. 13 (2013): 1230–38, http://archinte.jamanetwork.com/article.aspx?articleid=1710093.

14. Ibid.

15. William B. Grant, "Using Multicountry Ecological and Observational Studies to Determine Dietary Risk Factors for Alzheimer's Disease," *Journal of the American College of Nutrition* 35, no. 5 (July 2016): 476–89, http://www.tandfonline.com/doi/full/10.1080/07315724.2016.1161566.

16. Giovanni Vitale, Michael P. Brugts, Giulia Ogliari, et al., "Low Circulating IGF-I Bioactivity Is Associated with Human Longevity: Findings in Centenarians' Offspring," *Aging (Albany NY)* 4, no. 9 (September 2012): 580–89, https://www.ncbi.nlm.nih.gov/pmc/articles/PMC3492223/.

17. Rhonda Patrick, "The IGF-1 Trade-Off: Performance vs. Longevity," WellnessFX, September 4, 2013, http://blog.wellnessfx.com/2013/09/04/igf-1-trade-performance-vs-longevity/.

18. Fontana et al., "Long-Term Effects of Calorie or Protein Restriction."

19. M. F. McCarty, J. Barroso-Aranda, F. and Contreras, "The Low Methionine Content of Vegan Diets May Make

Methionine Restriction Feasible as a Life Extension Strategy," *Medical Hypotheses* 72 no. 2 (February 2009): 125–28, https://www.ncbi.nlm.nih.gov/pubmed/18789600/.

20. Brigham and Women's Hospital, "Eating More Protein May Not Benefit Older Men," ScienceDaily, April 2, 2018, www.sciencedaily.com/releases/2018/04/180402123241.htm.
21. Dan Pardi, "Does Protein Restriction Slow Aging? What about the Daniel Fast?," HumanOS, December 9, 2015, http://blog.dansplan.com/does-protein-restriction-and-fasting-slow-the-aging-process-better-aging-part-3/.
22. In Young Choi, Laura Piccio, Patra Childress, et al., "A Diet Mimicking Fasting Promotes Regeneration and Reduces Autoimmunity and Multiple Sclerosis Symptoms," *Cell Reports* 15 no. 10 (June 2016): 2136–46, http://www.cell.com/cell-reports/fulltext/S2211-1247(16)30576-9.
23. Thomas T. Samaras and Harold Elrick, "Height, Body Size and Longevity," *Acta Medica Okayama* 53, no. 4 (1999): 149–69, https://www.ncbi.nlm.nih.gov/pubmed/10488402.
24. D. D. Miller, "Economies of Scale," *Challenge* 33 (1990): 58–61.
25. Thomas T. Samaras and Harold Elrick, "Height, Body Size, and Longevity: Is Smaller Better for the Human Body?," *Western Journal of Medicine* 176, no. 3 (June 2002): 206–8, https://www.researchgate.net/publication/11355931_Height_body_size_and_longevity_Is_smaller_better_for_the_human_body.

26. M. Murata, "Secular Trends in Growth and Changes in Eating Patterns of Japanese Children," *American Journal of Clinical Nutrition* 72, no. 5 (November 2000): 1379–83, https://www.ncbi.nlm.nih.gov/pubmed/11063481.

27. T. T. Samaras and L. H. Storms, "Impact of Height and Weight on Life Span," *Bulletin of the World Health Organization* 70, no. 2 (1992): 259–67, https://www.ncbi.nlm.nih.gov/pubmed/1600586/.

28. Thomas T. Samaras, "How Body Height and Weight Affect Our Performance, Longevity, and Survival," *Journal of the Washington Academy of Sciences* 84, no. 3 (September 1996): 131–56.

29. D. Albanes, "Height, Early Energy Intake, and Cancer. Evidence Mounts for the Relation of Energy Intake to Adult Malignancies," *BMJ* 317, no. 7169 (November 1998): 1331–32, https://www.ncbi.nlm.nih.gov/pubmed/9812924/.

30. P. R. Hebert, U. Ajani, I. M. Lee, et al., "Adult Height and Incidence of Cancer in Male Physicians (United States)," *Cancer Causes Control* 8, no. 4 (July 1997): 591–97, https://www.ncbi.nlm.nih.gov/pubmed/9242474/.

31. J. Guevara-Aguirre, A. L. Rosenbloom, P. J. Fielder, et al., "Growth Hormone Receptor Deficiency in Ecuador: Clinical and Biochemical Phenotype in Two Populations," *Journal of Clinical Endocrinology and Metabolism* 76, no. 2 (February 1993): 417–23, https://www.ncbi.nlm.nih.gov/pubmed/7679400.

32. Adam Gesing, Khalid A. Al-Regaiey, Andrzej Bartke, et al., "Growth Hormone Abolishes Beneficial Effects of Calorie

Restriction in Long-Lived Ames Dwarf Mice," *Experimental Gerontology* 58 (August 2014): 219–29, https://www.researchgate.net/publication/265018637_Growth_Hormone_Abolishes_Beneficial_Effects_of_Calorie_Restriction_in_Long-Lived_Ames_Dwarf_Mice.

33. Isao Shimokawa, "Growth Hormone and IGF-1 Axis in Aging and Longevity," in Hormones in Ageing and Longevity, ed. S. Rattan and R. Sharma, vol. 6 (New York: Springer, 2017): 91–106, https://link.springer.com/chapter/10.1007%2F978-3-319-63001-4_5.

34. Kapil V. Ramachandran, Jack M. Fu, Thomas B. Schaffer, et al., "Activity-Dependent Degradation of the Nascentome by the Neuronal Membrane Proteasome," *Molecular Cell* 71, no. 1 (July 2018): 169–77, https://www.cell.com/molecular-cell/fulltext/S1097-2765(18)30455-6?_returnURL=https%3A%2F%2Flinkinghub.elsevier.com%2Fretrieve%2Fpii%2FS1097276518304556%3Fshowall%3Dtrue.

35. M. I. Frisard, A. Broussard, S. S. Davies, et al., "Aging, Resting Metabolic Rate, and Oxidative Damage: Results from the Louisiana Healthy Aging Study," *Journals of Gerontology* 62, no. 7 (July 2007): 752–59, https://www.ncbi.nlm.nih.gov/pubmed/17634323/.

36. Eun Jung Lee, Ji Young Kim, and Sang Ho Oh, "Advanced Glycation End Products (AGEs) Promote Melanogensis Through Receptor for AGEs," *Nature: Scientific Reports* 6 (June 2016): https://www.ncbi.nlm.nih.gov/pmc/articles/PMC4904211/.

37. F. J. Tessier, "The Maillard Reaction in the Human Body. The Main Discoveries and Factors That Affect Glycation," *Pathologie Biologie (Paris)* 58, no. 3 (June 2010): 214–19, https://www.ncbi.nlm.nih.gov/pubmed/19896783.
38. Sedsel Brøndum Lange, "Frequent Blood Donors Live Longer," ScienceNordic, November 20, 2015, http://sciencenordic.com/frequent-blood-donors-live-longer.
39. "Iron Accelerates Aging," World Health Net, December 30, 2015, https://www.worldhealth.net/news/iron-accelerates-aging/.
40. University of Wyoming, "Scientists Find Excess Mitochondrial Iron, Huntington's Disease Link," Science Daily, April 11, 2018, https://www.sciencedaily.com/releases/2018/04/180411145107.htm.
41. Barney E. Dwyer, Leo R. Zacharski, Dominic J. Balestra, et al., "Getting the Iron Out: Phlebotomy for Alzheimer's Disease?," *Medical Hypotheses* 72, no. 5 (2009): 504–9, https://www.ncbi.nlm.nih.gov/pmc/articles/PMC2732125/.
42. Nadja Schröder, Luciana Silva Figueiredo, and Maria Noêmia Martins de Lima, "Role of Brain Iron Accumulation in Cognitive Dysfunction: Evidence from Animal Models and Human Studies," *Journal of Alzheimer's Disease* 34, no. 4 (2013): 797–812, https://www.ncbi.nlm.nih.gov/pubmed/23271321.
43. Moataz Abdalkader, Riikka Lampinen, Katja M. Kanninen, et al., "Targeting Nrf2 to Supress Ferroptosis and Mitochondrial Dysfunction in Neurodegeneration,"

Frontiers in Neuroscience 12, no. 466 (July 2018): https://www.frontiersin.org/articles/10.3389/fnins.2018.00466/full.

44. K. P. Martinez, S. Eivaz-Mohammadi, and F. P. Gonzalex-Ibarra, "Effect of Phlebotomy on Motor UPDRS Score, Pain and Medication Dosage in a Patient with Parkinson's Disease and Hemochromatosis," *Archivos de Salud de Sinaloa* 6, no. 4 (2012): 118–20, http://hgculiacan.com/revistahgc/archivos/assin%2024%20aport_int2.pdf.

45. P. F. Silva, V. A. Garcia. S. Ada Dornelles, et al., "Memory Impairment Induced by Brain Iron Overload Is Accompanied by Reduced H3K9 Acetylation and Ameliorated by Sodium Butyrate," *Neuroscience* 200 (January 2012): 42–49, https://www.ncbi.nlm.nih.gov/m/pubmed/22067609/?i=3&from=/21593570/related.

46. Denise Minger, "The Truth about Ancel Keys: We've All Got it Wrong," Denise Minger (blog), December 22, 2011, https://deniseminger.com/2011/12/22/the-truth-about-ancel-keys-weve-all-got-it-wrong/.

47. Syed Haris Omar, "Oleuropein in Olive and Its Pharmacological Effects," *Scientia Pharmaceutica* 78, no. 2 (2010): 133–54, https://www.researchgate.net/publication/49703324_Oleuropein_in_Olive_and_its_Pharmacological_Effects.

48. "Consuming Extra Virgin Olive Oil Could Be a Viable Therapeutic Opportunity for Preventing or Halting Dementia and Alzheimer's Disease," BioFoundations, June 21, 2017, https://biofoundations.org/consuming-extra

-virgin-olive-oil-viable-therapeutic-opportunity-preventing-halting-dementia-alzheimers-disease/.

49. Charlotte Bamberger, Andreas Rossmeier, Katharina Lechner, et al., "A Walnut-Enriched Diet Affects Gut Microbiome in Healthy Caucasian Subjects: A Randomized, Controlled Trial," *Nutrients* 10, no. 2 (2018): 244, https://www.ncbi.nlm.nih.gov/pmc/articles/PMC5852820/.

50. H. D. Holscher, A. M. Taylor, K. S. Swanson, et al., "Almond Consumption and Processing Affects the Composition of the Gastrointestinal Microbiota of Healthy Adult Men and Women: A Randomized Controlled Trial," *Nutrients* 10, no. 2 (January 2018): https://www.ncbi.nlm.nih.gov/pubmed/29373513.

51. M. Ukhanova, X. Wang, D. J. Baer, et al., "Effects of Almond and Pistachio Consumption on Gut Microbiota Composition in Randomized Cross-Over Human Feeding Study," *British Journal of Nutrition* 111, no. 12 (June 2014): 2146–52, https://www.ncbi.nlm.nih.gov/pubmed/24642201.

52. Heather M. Guetterman, Kelly S. Swanson, Janet A. Novotny, et al., "Walnut Consumption Influences the Human Gut Microbiome," *FASEB Journal* 30, no. 1 (April 2016): http://www.fasebj.org/doi/abs/10.1096/fasebj.30.1_supplement.406.2.

53. Sebely Pal, Keith Woodford, Sonja Kukuljan, et al., "Milk Intolerance, Beta-Casein and Lactose," *Nutrients* 7, no. 9 (September 2015): 7285–97, http://www.ncbi.nlm.nih.gov/pmc/articles/PMC4586534/.

54. Keith Woodford, *Devil in the Milk: Illness, Health, and the Politics of A1 and A2 Milk* (White River Junction, VT: Chelsea Green Publishing Co., 2009).

Chapter 4: Get Younger from the Inside Out

1. Ryan P. Durk, Experanza Castillo, Leticia Márquez-Magaña, et al., "Gut Microbiota Composition Is Related to Cardiorespiratory Fitness in Healthy Young Adults," *Human Kinetics Journals* 0, no. 0 (2018): 1–15, https://journals.humankinetics.com/doi/10.1123/ijsnem.2018-0024.
2. Marion Tharrey, François Mariotti, Andrew Mashchak, et al., "Patterns of Plant and Animal Protein Intake Are Strongly Associated with Cardiovascular Mortality: the Adventist Health Study-2 Cohort," *International Journal of Epidemiology* 47, no. 5 (April 2010): https://academic.oup.com/ije/advance-article-abstract/doi/10.1093/ije/dyy030/4924399.
3. Medical College of Georgia at Augusta University, "Just One High-Fat Meal Sets the Perfect Stage for Heart Disease," ScienceDaily, March 29, 2018, https://www.sciencedaily.com/releases/2018/03/180329083259.htm.
4. University Hospitals Cleveland Medical Center, "Link Between Heart Attacks and Inflammatory Bowel Disease: Research Indicates Strong Role in Development of Cardiovascular Disease," ScienceDaily, March 6, 2018, https://www.sciencedaily.com/releases/2018/03/180306153734.htm.

5. Cristina Menni, Chihung Lin, Marina Cecelja, et al., "Gut Microbial Diversity Is Associated with Lower Arterial Stiffness in Women," *European Heart Journal* 39, no. 25 (July 2018): 2390–97, https://academic.oup.com/eurheartj/advance-article/doi/10.1093/eurheartj/ehy226/4993201.
6. C. Bogiatzi, G. Gloor, E. Allen-Vercoe, et al., "Metabolic Products of the Intestinal Microbiome and Extremes of Atherosclerosis," *Atherosclerosis* 273 (June 2018): 91–97, https://www.ncbi.nlm.nih.gov/pubmed/29702430.
7. Ancel Keys, "Human Atherosclerosis and the Diet," *AHAJournals* (2018): 115–18, http://circ.ahajournals.org/content/circulationaha/5/1/115.full.pdf.
8. W. X. Fan, R. Parker, B. Parpia, et al., "Erythrocyte Fatty Acids, Plasma Lipids, and Cardiovascular Disease in Rural China," *American Journal of Clinical Nutrition* 52, no. 6 (December 1990): 1027–36, https://www.ncbi.nlm.nih.gov/pubmed/2239777.
9. Tim Cutcliffe, "Omega-3 Intake Linked to Lower Risk of Death and Heart Disease: Framingham Data," Nutra Ingredients, March 20, 2018, https://www.nutraingredients.com/Article/2018/03/20/Omega-3-intake-linked-to-lower-risk-of-death-and-heart-disease-Framingham-data.
10. Uffe Ravnskov, David M. Diamond, and Rokura Hama, "Lack of an Association or an Inverse Association Between Low-Density-Lipoprotein Cholesterol and Mortality in the Elderly: A Systematic Review," *BMJ Journals* 6,

no. 6 (2016): 1–8, http://bmjopen.bmj.com/content/6/6/e010401.

11. Heiko Methe, Jong-Oh Kim, Sieglinde Kofler, et al., "Statins Decrease Toll-Like Receptor 4 Expression and Downstream Signaling in Human CD14+ Monocytes," *Arteriosclerosis, Thrombosis, and Vascular Biology* 25 (2005): 1439–45, https://www.ahajournals.org/doi/abs/10.1161/01.atv.0000168410.44722.86.

12. Jin Li, Shaoqiang Lin, Paul M. Vanhoutte, et al., "*Akkermansia Muciniphila* Protects Against Atherosclerosis by Preventing Metabolic Endotoxemia-Induced Inflammation in *Apoe* Mice," *Circulation* 133, no. 24 (April 2016): 2434–46, https://www.ahajournals.org/doi/abs/10.1161/circulationaha.115.019645.

13. Wenhua Zhu, Siwen Chen, Ronggui Chen, et al., "Taurine and Teal Polyphenols Combination Ameliorate Nonalcoholic Steatohepatitis in Rats," *BMC Complementary and Alternative Medicine* 17, no. 455 (2017): 1–12, https://bmccomplementalternmed.biomedcentral.com/track/pdf/10.1186/s12906-017-1961-3.

14. S. Lindeberg, P. Nilsson-Ehle, A. Terént, et al., "Cardiovascular Risk Factors in a Melanesian Population Apparently Free from Stroke and Ischaemic Heart Disease: The Kitava Study," *Journal of Internal Medicine* 236, no.3 (September 1994): 331–40, https://www.ncbi.nlm.nih.gov/pubmed/8077891.

15. Federica Cioffi, Rosalba Senese, Pasquale Lasala, et al., "Fructose-Rich Diet Affects Mitochondrial DNA Dam-

age and Repair in Rats," *Nutrients* 9, no. 4 (March 2017): 1–14, http://www.mdpi.com/2072-6643/9/4/323.

16. Richard Johnson, Santoz E. Perez-Pozo, Yuri Y. Sautin, et al., "Hypothesis: Could Excessive Fructose Intake and Uric Acid Cause Type 2 Diabetes?," *Endocrine Reviews* 30, no. 1 (February 2009): 96–116, https://www.researchgate.net/publication/23798203_Hypothesis_Could_Excessive_Fructose_Intake_and_Uric_Acid_Cause_Type_2_Diabetes.

17. Marek Kretowicz, Richard J. Johnson, Takuji Ishimoto, et al., "The Impact of Fructose on Renal Function and Blood Pressure," *International Journal of Nephrology* 2011 (2011): https://www.researchgate.net/publication/51524131_The_Impact_of_Fructose_on_Renal_Function_and_Blood_Pressure.

18. Y. Ilan, "Leaky Gut and the Liver: A Role for Bacterial Translocation in Nonalcoholic Steatohepatitis," *World Journal of Gastroenterology* 18, no. 21 (June 7, 2012): 2609–18, https://www.ncbi.nlm.nih.gov/pubmed/22690069.

19. A-Sol Kim and Hae-Jin Ko, "Plasma Concentrations of Zonulin Are Elevated in Obese Men with Fatty Liver Disease," *Diabetes, Metabolic Syndrome and Obesity: Targets and Therapy* 11 (October 2018): 149–57, https://www.dovepress.com/getfile.php?fileID=41677.

20. Tufts University, "Gut Check: Metabolites Shed by Intestinal Microbiota Keep Inflammation at Bay: Researchers Find Inflammatory Response in Fatty Liver Disease Is Reduced by Two Tryptophan Metabolites from Gut Bac-

teria," ScienceDaily, May 4, 2018, https://www.sciencedaily.com/releases/2018/05/180504103743.htm.

21. Brigham and Women's Hospital, "Healthy Diet May Lower Risk of Hearing Loss in Women: Patterns of Healthy Eating May Lower Risk of Hearing Loss by 30 Percent," ScienceDaily, May 11, 2018, https://www.sciencedaily.com/releases/2018/05/180511123022.htm.

22. University of Pennsylvania School of Medicine, "Potential of Manipulating Gut Microbiome to Boost Efficacy of Cancer Immunotherapies," ScienceDaily, April 2, 2018, https://www.sciencedaily.com/releases/2018/04/180402171038.htm.

23. NIH/National Cancer Institute, "Gut Microbiome Can Control Antitumor Immune Function in Liver," ScienceDaily, May 24, 2018, https://www.sciencedaily.com/releases/2018/05/180524141715.htm.

24. Smruti Pushalkar, Mautin Hundeyin, Donnele Daley, et al., "The Pancreatic Cancer Microbiome Promotes Oncogenesis by Induction of Innate and Adaptive Immune Suppression," *American Association for Cancer Research* 8, no. 4 (2018): 1–14, http://cancerdiscovery.aacrjournals.org/content/early/2018/03/08/2159-8290.CD-17-1134.

25. University of Leeds, "Links Between Eating Red Meat and Distal Colon Cancer in Women," ScienceDaily, April 2, 2018, https://www.sciencedaily.com/releases/2018/04/180402085853.htm.

26. Matthew G. Vander Heiden, Lewis C. Cantley, and Craig B. Thompson, "Understanding the Warburg Effect: The

Metabolic Requirements of Cell Proliferation," *Science* 324, no. 5930 (May 2009): 1029–33, http://www.ncbi.nlm.nih.gov/pmc/articles/PMC2849637/.

27. Ibid.

28. Mitch Leslie, "Putting Immune Cells on a Diet," *Science* 359, no. 6383 (March 2018): 1454–56, http://science.sciencemag.org/content/359/6383/1454.full.

29. Duke University, "Metastatic Cancer Gorges on Fructose in the Liver: Discovery of Metabolic Reprogramming in Metastatic Cancer Could Lead to New Therapies," ScienceDaily, April 26, 2018, https://www.sciencedaily.com/releases/2018/04/180426141533.htm.

30. Hokkaido University, "Obesity Inhibits Key Cancer Defense Mechanism: Obesity Could Enhance Cancer Development While Aspirin Might Prevent It—a New Insight into Potential Targets for Cancer Prevention," ScienceDaily, April 26, 2018, https://www.sciencedaily.com/releases/2018/04/180426102844.htm.

31. Temidayo Fadelu, Donna Niedzwiecki, Sui Zhang, et al., "Nut Consumption and Survival in Patients with Stage III Colon Cancer: Results from CALGB 89803 (Alliance)," *Journal of Clinical Oncology* 36, no. 11 (April 10, 2018), https://www.scholars.northwestern.edu/en/publications/nut-consumption-and-survival-in-patients-with-stage-iii-colon-can.

32. R. Singh, S. Subramanian, J. M. Rhodes, et al., "Peanut Lectin Stimulates Proliferation of Colon Cancer Cells by

Interaction with Glycosylated CD44v6 Isoforms and Consequential Activation of c-Met and MAPK: Functional Implications for Disease-Associated Glycosylation Changes," *Glycobiology* 16, no. 7 (July 2006): 594–601, https://www.ncbi.nlm.nih.gov/pubmed/16571666.

33. Masako Nakanishi, Yanfei Chen, Veneta Qendro, et al., "Effects of Walnut Consumption on Colon Carcinogenesis and Microbial Community Structure," *Cancer Prevention Research* 9, no. 8 (August 2016): 692–703, http://cancerpreventionresearch.aacrjournals.org/content/9/8/692.

34. Jennifer T. Lee, Gabriel Y. Lai, Linda M. Liao, et al., "Nut Consumption and Lung Cancer Risk: Results from Two Large Observational Studies," *Cancer Epidemiology Biomarkers and Prevention* 26, no. 6 (June 26, 2017): 826–36, https://www.ncbi.nlm.nih.gov/pubmed/28077426.

35. G. Grosso, J. Yang, S. Marventano, et al., "Nut Consumption on All-Cause, Cardiovascular, and Cancer Mortality Risk: A Systematic Review and Meta-Analysis of Epidemiologic Studies," *American Journal of Clinical Nutrition* 101, no. 4 (April 2015): 783–93, https://www.ncbi.nlm.nih.gov/pubmed/25833976/.

36. B. Gopinath, V. M. Flood, G. Burlutksy, et al., "Consumption of Nuts and Risk of Total and Cause-Specific Mortality over 15 Years," *Nutrition, Metabolism, and Cardiovascular Disease* 25, no. 12 (December 2015): 1125–1231, https://www.ncbi.nlm.nih.gov/pubmed/26607701/.

Chapter 5: Dance Your Way into Old Age

1. University of Rochester Medical Center, "The Bugs in Your Gut Could Make You Weak in the Knees: A Prebiotic May Alter the Obese Microbiome and Protect Against Osteoarthritis," ScienceDaily, April 19, 2018, https://www.sciencedaily.com/releases/2018/04/180419100135.htm.
2. P. Fritz, H. V. Tuczek, J. Hoenes, et al. "Use of Lectin-Immunohistochemistry in Joint Diseases," *Acta Histochemica* 36 (1998): 277–83, https://www.ncbi.nlm.nih.gov/pubmed/3150561/.
3. S. Yoshino, E. Sasatomi, and M. Ohsawa, "Bacterial Lipopolysaccharide Acts as an Adjuvant to Induce Autoimmune Arthritis in Mice," *Immunology* 99, no. 4 (2000): 607–14, https://www.ncbi.nlm.nih.gov/pmc/articles/PMC2327198/.
4. Michael L. Mishkind, Barry A. Palevitz, Natasha V. Raikhel, et al., "Localization of Wheat Germ Agglutinin–like Lectins in Various Species of the Gramineae," *Science* 220, no. 4603 (June 17, 1983): 1290–92, http://science.sciencemag.org/content/220/4603/1290.
5. Laurent Léotoing, Marie-Jeanne Davicco, Patrice Lebecque, et al., "The Flavonoid Fisetin Promotes Osteoblasts Differentiation Through Runx2 Transcriptional Activity," *Molecular Nutrition and Food Research* 58, no. 6 (February 2014): https://onlinelibrary.wiley.com/doi/abs/10.1002/mnfr.201300836.

6. K. E. Brickett, J. P.Dahiya, H. L. Classen, et al., "The Impact of Nutrient Density, Feed Form, and Photoperiod on the Walking Ability and Skeletal Quality of Broiler Chickens," *Poultry Science* 86, no. 10 (October 2007): 2117–25, http://ps.oxfordjournals.org/content/86/10/2117.full.
7. L. Jakobsen, P. Garneau, G. Bruant, et al., "Is *Escherichia coli* Urinary Tract Infection a Zoonosis? Proof of Direct Link with Production Animals and Meat," *European Journal of Clinical Microbiology and Infectious Diseases* 31, no. 6 (October 2011): 1121–29, http://www.ncbi.nlm.nih.gov/m/pubmed/22033854/.
8. The Endocrine Society, "Mediterranean Diet Is Linked to Higher Muscle Mass, Bone Density after Menopause," ScienceDaily, March 18, 2018, https://www.sciencedaily.com/releases/2018/03/180318144826.htm.
9. University at Buffalo, "Strenuous Exercise in Adolescence May Ward Off Height Loss Later in Life: Researchers Identified Several Key Factors in Postmenopausal Women Associated with Marked Height Loss of More Than 1 Inch," ScienceDaily, May 23, 2018, https://www.sciencedaily.com/releases/2018/05/180523133405.htm.
10. Westmead Institute for Medical Research, "Exercise Cuts Risk of Chronic Disease in Older Adults," Science Daily, July 23, 2018, https://www.sciencedaily.com/releases/2018/07/180723142920.htm.
11. Queensland University of Technology, "Older People Advised to Dance for Better Posture, Flexibility, Energy

and Happiness," ScienceDaily, April 5, 2018, https://www.sciencedaily.com/releases/2018/04/180405093254.htm.

12. V. Billat, G. Dhonneur, L. Mille-Hamard, et al., "Case Studies in Physiology: Maximal Oxygen Consumption and Performance in a Centenarian Cyclist," *Journal of Applied Physiology* 122, no. 3 (March 2017): 430–34, https://www.ncbi.nlm.nih.gov/pubmed/28035015.

13. J. Hentilä, J. P. Ahtiainnen, G. Paulsen, et al., "Autophagy Is Induced by Resistance Exercise in Young Men, but Unfolded Protein Response Is Induced Regardless of Age," *Acta Physiologica* 224, no. 1 (April 2018): https://onlinelibrary.wiley.com/doi/abs/10.1111/apha.13069.

14. Harvard University, "Exercise Could Make the Heart Younger: Mice Make over Four Times as Many New Heart Muscle Cells When They Exercise, Study Finds," Science Daily, April 25, 2018, https://www.sciencedaily.com/releases/2018/04/180425093804.htm.

15. American Academy of Neurology, "Physically Fit Women Nearly 90 Percent Less Likely to Develop Dementia," ScienceDaily, March 15, 2018, https://www.sciencedaily.com/releases/2018/03/180315101805.htm.

16. Jill K. Morris, Eric D. Vidoni, David K. Johnson, et al., "Aerobic Exercise for Alzheimer's Disease: A Randomized Controlled Pilot Trial," *PLOS One* 10 (February 2017): 1–14, https://www.drperlmutter.com/wp-content/uploads/2018/01/Aerobic-exercise-for-Alzheimers-disease-A-randomized-controlled-pilot-trial.pdf.

17. Frontiers, "Leg Exercise Is Critical to Brain and Nervous System Health," ScienceDaily, May 23, 2018, https://www.sciencedaily.com/releases/2018/05/180523080214.htm.

18. University of Western Ontario, "Brain Game Doesn't Offer Brain Gain," ScienceDaily, July 30, 2018, https://www.sciencedaily.com/releases/2018/07/180730120405.htm.

19. L. L. Ji, M. C. Gomez-Cabrera, and J. Vina, "Exercise and Hormesis: Activation of Cellular Antioxidant Signaling Pathway," *Annals of the New York Academy of Sciences* 1067 (May 2006): 425–35, https://www.ncbi.nlm.nih.gov/m/pubmed/16804022/?i=5&from=/16906627/related.

20. Jingyuan Chen, Yuan Guo, Yajun Gui, et al., "Physical Exercise, Gut, Gut Microbiota, and Atherosclerotic Cardiovascular Diseases," *Lipids in Health and Disease* 17, no. 17 (January 2018), https://lipidworld.biomedcentral.com/articles/10.1186/s12944-017-0653-9.

21. Siobhan F. Clarke, Eileen F. Murphy, Orla O'Sullivan, et al., "Exercise and Associated Dietary Extremes Impact on Gut Microbial Diversity," *Gut* 0 (June 2014): 1–8, http://www.natap.org/2014/HIV/Gut-2014-Clarke-gutjnl-2013-306541.pdf.

22. Jeong June Choi, Sung Yong Eum, Evadnie Rampersaud, et al., "Exercise Attenuates PCB-Induced Changes in the Mouse Gut Microbiome," *Environmental Health Perspectives* 121, no. 6 (April 2013): 725–30, https://www

.researchgate.net/publication/236598953_Exercise_Attenuates_PCB-Induced_Changes_in_the_Mouse_Gut_Microbiome; Janet Chow, Haiqing Tang, and Sarkis K. Mazmanian, "Pathobionts of Gastrointestinal Microbiota and Inflammatory Disease," *Science Direct: Current Opinion on Immunology*, 23, no. 4 (August 2014): 273–480, https://www.sciencedirect.com/science/article/pii/S0952791511000835.

23. Ag Cox, D. B. Pyne, P. U. Saunders, et al., "Oral Administration of the Probiotic Lactobacillus Fermentum VRI-003 and Mucosal Immunity in Endurance Athletes," *British Journal of Sports Medicine* 44, no. 4 (March 2010): 222–26, https://www.ncbi.nlm.nih.gov/pubmed/18272539.

24. M. Matsumoto, R. Inoue, and T. Tsukahara, "Voluntary Running Exercise Alters Microbiota Composition and Increases N-Butyrate Concentration in the Rat Cecum," *Bioscience, Biotechnology, and Biochemistry* 72, no. 2 (February 2008): 572–76, https://www.ncbi.nlm.nih.gov/pubmed/18256465.

25. Queensland University of Technology, "Older People Advised to Dance for Better Posture, Flexibility, Energy and Happiness."

26. Harvard Health Publishing, "Exercise Is an All-Natural Treatment to Fight Depression," Harvard Medical School, April 30, 2018, https://www.health.harvard.edu/mind-and-mood/exercise-is-an-all-natural-treatment-to-fight-depression.

27. Siri Carpenter, "That Gut Feeling," *American Psycho-*

logical Association 43, no. 8 (September 2012): 50, http://www.apa.org/monitor/2012/09/gut-feeling.aspx.

28. Herman Pontzer, "The Exercise Paradox," *Scientific American*, February 2017, https://www.scientificamerican.com/article/the-exercise-paradox/.

29. David C. Nieman, "Marathon Training and Immune Function," *Sports Medicine* 37, nos. 4–5 (April 2007): 412–15, https://link.springer.com/article/10.2165/00007256-200737040-00036.

30. James H. O'Keefe, Harshal R. Patil, Carl J. Lavie, et al., "Potential Adverse Cardiovascular Effects from Excessive Endurance Exercise," *Mayo Clinic Proceedings* 87, no. 6 (June 2012): 587–95, https://www.sciencedirect.com/science/article/pii/S0025619612004739.

31. M. C. Gomez-Cabrera, A. Martínez, G. Santangelo, et al., "Oxidative Stress in Marathon Runners: Interest of Antioxidant Supplementation," *British Journal of Nutrition* 96, no. 1 (2006): 1–3, https://www.ncbi.nlm.nih.gov/m/pubmed/16923247/?i=3&from=/18191748/related.

32. M. C. Gomez-Cabrera, E. Domenech, and J. Viña, "Moderate Exercise Is an Antioxidant: Upregulation of Antioxidant Genes by Training," *Free Radical Biology and Medicine* 44, no. 2 (2008): 126–31, https://www.ncbi.nlm.nih.gov/m/pubmed/18191748/?i=3&from=/16804022/related.

33. Katrin Gutekunst, Karsten Krüger, Christian August, et al., "Acute Exercises Induce Disorders of the Gastrointestinal Integrity in a Murine Model," *European Journal* of *Applied Physiology* 114, no. 3 (March 2014): 609–17, https://www.ncbi.nlm.nih.gov/pubmed/24352573.

Chapter 6: Remember Your Old Age

1. Maura Boldrini, Camille A. Fulmore, Alexandria N. Tartt, et al., "Human Hippocampal Neurogenesis Persists Throughout Aging," *Cell Stem Cell* 22, no. 4 (April 2018): 589–99, https://www.cell.com/cell-stem-cell/references/S1934-5909(18)30121-8.
2. K. Segaert, S. J. E. Lucas, C. V. Burley, et al., "Higher Physical Fitness Levels Are Associated with Less Language Decline in Healthy Ageing," *Scientific Reports* 8, no. 6715 (2018): https://www.nature.com/articles/s41598-018-24972-1.
3. Steven R. Gundry, "Abstract P238: Remission/Cure of Autoimmune Diseases by a Lectin Limite Diet Supplemented with Probiotics, Prebiotics, and Polyphenols," *Circulation* 137, no. 1 (June 2018): http://circ.ahajournals.org/content/137/Suppl_1/AP238.
4. Karen A. Scott, Masayuki Ida, Veronica L. Peterson, et al., "Revisiting Metchnikoff: Age-Related Alterations in Microbiota-Gut-Brain Axis in the Mouse," *Brain, Behavior, and Immunity* 65 (October 2017): 20–32, https://www.sciencedirect.com/science/article/pii/S088915911730034X.
5. Annamaria Cattaneo, Nadia Cattane, Samantha Galluzzi, et al., "Association of Brain Amyloidosis with

Pro-Inflammatory Gut Bacterial Taxa and Peripheral Inflammation Markers in Cognitively Impaired Elderly," *Neurobiology of Aging* 49 (January 2017): 60–68, https://www.sciencedirect.com/science/article/pii/S0197458016 30197X.

6. Gian D. Pal, Maliha Shaikh, Christopher B. Forsyth, et al., "Abnormal Lipopolysaccharide Binding Protein as Marker of Gastrointestinal Inflammation in Parkinson Disease," *Frontiers in Neuroscience* 9, no. 306 (September 2015): https://www.ncbi.nlm.nih.gov/pmc/articles /PMC4555963/.
7. University of Erlangen-Nuremberg, "Aggressive Immune Cells Aggravate Parkinson's Disease," ScienceDaily, July 19, 2018, https://www.sciencedaily.com/releases/2018 /07/180719094349.htm.
8. Jolene Zheng, Mingming Wang, Wenqian Wei, et al., "Dietary Plant Lectins Appear to Be Transported from the Gut to Gain Access to and Alter Dopaminergic Neurons of *Caenorhabditis elegans*, a Potential Etiology of Parkinson's Disease," *Frontiers in Neurosciences* 3, no. 7 (March 2016): http://journal.frontiersin.org/article/10.3389/fnut.2016 .00007/full.
9. Ibid.
10. Bojing Liu, Fang Fang, Nancy L. Pedersen, et al., "Vagotomy and Parkinson Disease: A Swedish Register–Based Matched-Cohort Study," *Neurology* 88, no. 21 (May 23, 2017): 1996–2002, https://www.ncbi.nlm.nih.gov/pubmed /28446653.

11. Thibaud Lebouvier, Michel Neunlist, Emmanuel Coron, et al., "Colonic Biopsies to Assess the Neuropathology of Parkinson's Disease and Its Relationship with Symptoms," *PLOS One* 5, no. 9 (September 2010): https://www.researchgate.net/profile/Michel_Neunlist/publication/46382378_Colonic_Biopsies_to_Assess_the_Neuropathology_of_Parkinson%27s_Disease_and_Its_Relationship_with_Symptoms/links/02bfe50f5c59b2fc7a000000/Colonic-Biopsies-to-Assess-the-Neuropathology-of-Parkinsons-Disease-and-Its-Relationship-with-Symptoms.pdf?origin=publication_detail.

12. Florida Atlantic University, "Mutation of Worm Gene, swip-10, Triggers Age-Dependent Death of Dopamine Neurons: Death of Dopamine-Producing Cells Key Feature of Parkinson's Disease," ScienceDaily, April 4, 2018, https://www.sciencedaily.com/releases/2018/04/180404093926.htm.

13. Gary W. Small, Prabha Siddarth, Zhaoping Li, et al., "Memory and Brain Amyloid and Tau Effects of a Bioavailable Form of Curcumin in Non-Demented Adults: A Double-Blind, Placebo-Controlled 18-Month Trial," *American Journal of Geriatric Psychiatry* 26, no. 3 (March 2018): 266–77, https://www.sciencedirect.com/science/article/pii/S1064748117305110.

14. Y. Wang, S. Begum-Haque, K. M. Telesford, et al., "A Commensal Bacterial Product Elicits and Modulates Migratory Capacity of CD39(+) CD4 T Regulatory Subsets in the Suppression of Neuorinflammation," *Gut Microbes*

5, no. 4 (July 2014): 552–61, https://www.ncbi.nlm.nih.gov/m/pubmed/25006655/?i=4&from=/20817872/related.

15. J. Ochoa-Repáraz, D. W. Mielcarz, L. E. Ditrio, et al., "Central Nervous System Demyelinating Disease Protection by the Human Commensal *Bacteroides fragilis* Depends on Polysaccharide A Expression," *Journal of Immunology* 185, no. 7 (October 2010): 4101–8, https://www.ncbi.nlm.nih.gov/m/pubmed/20817872/.
16. Gundry, "Abstract P238: Remission/Cure of Autoimmune Diseases by a Lectin Limite Diet Supplemented with PRobiogics, Prebiotics and Polyphenols."
17. C. Jiang, G. Li, Z. Liu, et al., "The Gut Microbiota and Alzheimer's Disease," *Journal of Alzheimer's Disease* 58, no. 1 (2017): 1–15, https://www.ncbi.nlm.nih.gov/pubmed/28372330/.
18. Yuhai Zhao and Walter J. Lukiw, "Microbiome-Generated Amyloid and Potential Impact on Amyloidogenesis in Alzheimer's Disease (AD)," *Journal of Nature Science* 1, no. 7 (July 2015): 138, http://europepmc.org/articles/PMC4469284/.
19. Lulu Xie, Hongyi Kang, Qiwu Xu, et al., "Sleep Drives Metabolite Clearance from the Adult Brain," *Science* 342, no. 6156 (October 2014): https://www.ncbi.nlm.nih.gov/pmc/articles/PMC3880190/.
20. University College London, "Obesity Increases Dementia Risk," ScienceDaily, November 30, 2017, https://www.sciencedaily.com/releases/2017/11/171130133812.htm.

21. María-Isabel Covas, Montserrat Fitó, Jaume Marrugat, et al., "The Effect of Polyphenols in Olive Oil on Heart Disease Risk Factors: A Randomized Trial," *Annals of Internal Medicine* 145, no. 5 (September 2006): http://annals.org/aim/article-abstract/727945/effect-polyphenols-olive-oil-heart-disease-risk-factors-randomized-trial.
22. "Consuming Extra Virgin Olive Oil Could Be a Viable Therapeutic Opportunity for Preventing or Halting Dementia and Alzheimer's Disease," BioFoundations, June 21, 2017, https://biofoundations.org/consuming-extra-virgin-olive-oil-viable-therapeutic-opportunity-preventing-halting-dementia-alzheimers-disease/.
23. Vanessa Pitozzi, Michela Jacomelli, Dolores Catelan, et al., "Long-Term Dietary Extra-Virgin Olive Oil Rich in Polyphenols Reverses Age-Related Dysfunctions in Motor Coordination and Contextual Memory in Mice: Role of Oxidative Stress," *Rejuvenation Research* 15, no. 6 (January 2013): https://www.liebertpub.com/doi/abs/10.1089/rej.2012.1346.
24. Jedha Dening, "Olive Oil Component Stops Gut Bacteria Linked to Heart and Brain Diseases," Olive Oil Times, February 16, 2016, https://www.oliveoiltimes.com/olive-oil-health-news/olive-oil-stops-gut-bacteria-linked-to-heart-and-brain-diseases/50507.
25. Valentina Carito, Mauro Ceccanti, George Chaldakov, et al., "Polyphenols, Nerve Growth Factor, Brain-Derived Neurotrophic Factor, and the Brain," in *Bioactive Nutraceuticals and Dietary Supplements in Neurological and*

Brain Disease, ed. Ronald Watson and Victor Preedy (New York: Academic Press, 2015), 65–71, https://www.researchgate.net/profile/George_Chaldakov/publication/312041499_NGF_BDNF_olive_oil_polyphnols/links/586c0b1008aebf17d3a5b3a6/NGF-BDNF-olive-oil-polyphnols.pdf.

26. P. G. Prieto, J. Cancelas, M. L. Villanueva-Peñacarrillo, et al., "Effects of an Olive-Oil-Enriched Diet on Plasma GLP-1 Concentration and Intestinal Content, Plasma Insulin Concentration, and Glucose Tolerance in Normal Rats," *Endocrine* 26, no. 2 (March 2005): 107–15, https://www.ncbi.nlm.nih.gov/pubmed/15888922.

27. A. M. Bak, L. Egefjord, M. Gejl, et al., "Targeting Amyloid-Beta by Glucagon-Like Peptide-1 (GLP-1) in Alzheimer's Disease and Diabetes," *Expert Opinions on Therapeutic Targets* 15, no. 10 (October 2011): 1153–52, https://www.ncbi.nlm.nih.gov/pubmed/21749267/.

28. Sara De Nicoló, Luigi Tarani, Mauro Ceccanti, et al., "Effects of Olive Polyphenols Administration on Nerve Growth Factor and Brain-Derived Neurotrophic Factor in the Mouse Brain," *Nutrition* 29, no. 4 (April 2013): 681–87, https://www.sciencedirect.com/science/article/pii/S0899900712004303.

29. Lisa Rapaport, "Mediterranean Diet with Olive Oil, Nuts Linked to Healthier Brain," *Scientific American*, 2018, https://www.scientificamerican.com/article/mediterranean-diet-with-olive-oil-nuts-linked-to-healthier-brain/.

30. Ravinder Nagpal, Carol A. Shively, Susan A. Appt, et al., "Gut Microbiome Composition in Non-Human Primates Consuming a Western or Mediterranean Diet," *Frontiers in Nutrition* (April 2018): https://www.frontiersin.org/articles/10.3389/fnut.2018.00028/full.

31. Michelle Luciano, Janie Corely, Simon R. Cox, et al., "Mediterranean-Type Diet and Brain Structural Change From 73 to 76 Years in a Scottish Cohort," *Neurology* (January 2017): http://n.neurology.org/content/early/2017/01/04/WNL.0000000000003559.

32. Y. Zhang, P. Zhuang, W. He, et al., "Association of Fish and Long-Chain Omega-3 Fatty Acids Intakes with Total and Cause-Specific Mortality: Prospective Analysis of 421 309 Individuals," *Journal of Internal Medicine* 284, no. 4 (July 2018): https://onlinelibrary.wiley.com/doi/abs/10.1111/joim.12786.

33. IOS Press, "Can Omega-3 Help Prevent Alzheimer's Disease? Brain SPECT Imaging Shows Possible Link," ScienceDaily, May 19, 2017, https://www.sciencedaily.com/releases/2017/05/170519124034.htm.

34. James V. Pottala, Kristine Yaffe, Jennifer G. Robinson, et al., "Higher RBC EPA+ DHA Corresponds with Larger Total Brain and Hippocampal Volumes," *Neurology* (January 2014), http://n.neurology.org/content/early/2014/01/22/WNL.0000000000000080.short.

35. Martha Clare Morris, Yamin Wang, Lisa L. Barnes, et al., "Nutrients and Bioactives in Green Leafy Vegetables and Cognitive Decline," *Neurology* 90, no. 3 (January 16,

2018): e214–e222, http://n.neurology.org/content/90/3/e214.

36. S. F. Clarke, E. F. Murphy, O. O'Sullivan, et al., "Exercise and Associated Dietary Extremes Impact on Gut Microbial Diversity," *Gut* 63, no. 12 (June 2014): 1913–20, https://www.ncbi.nlm.nih.gov/m/pubmed/25021423/.

37. S. R. Knowles, E. A. Nelson, E. A. Palombo, "Investigating the Role of Perceived Stress on Bacterial Flora Activity and Salivary Cortisol Secretion: A Possible Mechanism Underlying Susceptibility to Illness," *Biological Psychiatry* 77, no. 2 (February 2008): 132–37, https://www.ncbi.nlm.nih.gov/m/pubmed/18023961/.

38. Monash University, "Extreme Exercise Linked to Blood Poisoning," ScienceDaily, June 16, 2015, https://www.sciencedaily.com/releases/2015/06/150616093646.htm.

39. N. Mach, Y. Ramayo-Caldas, et al., "Understanding the Response to Endurance Exercise Using a Systems Biology Approach: Combining Blood Metabolomics, Transcriptomics and miRNomics in Horses," *BMC Genomics* 18, no. 1 (February 2017): ncbi.nlm.nih.gov/pubmed/28212624.

40. Hong-Li Li, Lan Lu, Xiao-Shuang Wang, et al.,"Alteration of Gut Microbiota and Inflammatory Cytokine/Chemokine Profiles in 5-Fluorouracil Induced Intestinal Mucositis," *Frontiers in Cellular and Infection Microbiology* 7 (October 26, 2017): 455, https://www.frontiersin.org/articles/10.3389/fcimb.2017.00455/full.

41. Shadi S. Yarandi, Daniel A. Peterson, Glen J. Treisman, et

al., "Modulatory Effects of Gut Microbiota on the Central Nervous System: How Gut Could Play a Role in Neuropsychiatric Health and Diseases," *Neurogastroenterology and Motility* 22, no. 2 (April 2016): 201–12, https://www.ncbi.nlm.nih.gov/pmc/articles/PMC4819858/.

42. Timothy G. Dinan, Roman M. Stilling, Catherine Stanton, et al., "Collective Unconscious: How Gut Microbes Shape Human Behavior," *Journal of Psychiatric Research* 63 (April 2015): 1–9, https://www.sciencedirect.com/science/article/pii/S0022395615000655#bib21.

43. Stephen M. Collins, Zain Kassam, and Premysl Bercik, "The Adoptive Transfer of Behavioral Phenotype via the Intestinal Microbiota: Experimental Evidence and Clinical Implications," *Current Opinion in Microbiology* 16, no. 3 (June 2013): 240–45, https://www.sciencedirect.com/science/article/pii/S1369527413000787.

44. Dinan et al., "Collective Unconscious."

45. Maia Szalavitz, "Explaining Why Meditators May Live Longer," *Time*, Body and Mind, December 23, 2010, http://healthland.time.com/2010/12/23/could-meditation-extend-life-intriguing-possibility-raised-by-new-study/.

46. Sanchari Sinha, Som Hath Singh, Y. P. Monga, et al., "Improvement of Glutathione and Total Antioxidant Status with Yoga," *Journal of Alternative and Complementary Medicine* 13, no. 10 (December 2007): https://www.liebertpub.com/doi/abs/10.1089/acm.2007.0567.

47. Henrike M. Hamer, Daisy M. A. E. Jonkers, Aalt Bast, et al., "Butyrate Modulates Oxidative Stress in the Colonic Muscosa of Healthy Humans," *Clinical Nutrition* 23, no. 1 (February 2009): 88–93, https://www.sciencedirect.com/science/article/pii/S0261561408002227.

48. Rameswar Pal, Som Nath Singh, Abhirup Chatterjee, et al., "Age-Related Changes in Cardiovascular System, Autonomic Functions, and Levels of BDNF of Healthy Active Males: Role of Yogic Practice," *Age* 36, no. 9683 (July 2014): https://link.springer.com/article/10.1007/s11357-014-9683-7.

49. Raeesah Maqsood and Trevor W. Stone, "The Gut-Brain Axis, BDNF, NMDA and CNS Disorders," *Neurochemical Research* 41, no. 11 (November 2016): 2819–35, https://link.springer.com/article/10.1007/s11064-016-2039-1.

50. Brett Froeliger, Eric L. Garland, and F. Joseph McClernon, "Yoga Meditation Practitioners Exhibit Greater Gray Matter Volume and Fewer Reported Cognitive Failures: Results of a Preliminary Voxel-Based Morphometric Analysis," *Hindawi* (October 2012): https://www.hindawi.com/journals/ecam/2012/821307/.

51. Trinity College Dublin, "The Yogi Masters Were Right—Meditation and Breathing Exercises Can Sharpen Your Mind: New Research Explains Link Between Breath-Focused Meditation and Attention and Brain Health," ScienceDaily, May 10, 2018, https://www.sciencedaily.com/releases/2018/05/180510101254.htm.

52. Harris A. Eyre, Bianca Acevedo, Hongyu Yang, et al., "Changes in Neural Connectivity and Memory Following a Yoga Intervention for Older Adults: A Pilot Study,"

Journal of Alzheimer's Disease 52, no. 2 (May 2016): 673–84, https://content.iospress.com/articles/journal-of-alzheimers-disease/jad150653.

53. Massachusetts General Hospital, "How Exercise Generates New Neurons, Improves Cognition in Alzheimer's Mouse: How to Mimic the Beneficial Effects of Exercise," ScienceDaily, September 6, 2018, https://www.sciencedaily.com/releases/2018/09/180906141623.htm.

Chapter 7: Look Younger As You Age

1. Jan Gruber and Brian K. Kennedy, "Microbiome and Longevity: Gut Microbes Send Signals to Host Mitochondria," *Cell* 169, no. 7 (June 2017): 1168–69, http://www.cell.com/cell/fulltext/S0092-8674(17)30641-4.
2. University of Chicago Medical Center, "Specific Bacteria in the Small Intestine Are Crucial for Fat Absorption: A High-Fat Diet Promotes Growth of the Microbes That Boost Lipid Digestion and Absorption," ScienceDaily, April 11, 2018, https://www.sciencedaily.com/releases/2018/04/180411131639.htm.
3. A. Janesick and B. Blumberg, "Endocrine Disrupting Chemicals and the Developmental Programming of Adipogenesis and Obesity," *Birth Defects Research Part C: Embryo Today* 93, no. 1 (March 2011): 34–50, http://www.ncbi.nlm.nih.gov/m/pubmed/21425440/.
4. Gretchen Goldman, Christina Carlson, Yixuan Zhang, *Bad Chemistry: How the Chemical Industry's Trade Associa-*

tion Undermines the Policies That Protect Us (Cambridge, MA: Center for Science and Democracy, July 2015), http://www.ucsusa.org/center-science-and-democracy/fighting-misinformation/american-chemistry-council-report#.WBpRlNw6NcB.

5. Brian Bienkowski, "BPA Replacement Also Alters Hormones," *Scientific American*, January 17, 2013, https://www.scientificamerican.com/article/bpa-replacement-also-alters-hormones/.
6. P. M. D. Foster, R. C. Cattley, and E. Mylchreest, "Effects of Di-*n*-Butyl Phthalate (DBP) on Male Reproductive Development in the Rat: Implications for Human Risk Assessment," *Food and Chemical Toxicology* 38 (suppl. 1) (April 1, 2000): S97–S99, https://www.sciencedirect.com/science/article/pii/S0278691599001283.
7. Susan M. Duty, Narendra P. Singh, Manori J. Silva, et al., "The Relationship Between Environmental Exposures to Phthalates and DNA Damage in Human Sperm Using the Neutral Comet Assay," *Environmental Health Perspectives* 111, no. 9 (July 2003): 1164–69.
8. Ivelisse Colón, Doris Caro, Carlos J. Bourdony, et al., "Identification of Phthalate Esters in the Serum of Young Puerto Rican Girls with Premature Breast Development," *Environmental Health Perspectives* 108, no. 9 (September 2000): 895–900, https://www.researchgate.net/publication/12305674_Identification_of_Phthalate_Esters_in_the_Serum_of_Young_Puerto_Rican_Girls_with_Premature_Breast_Development.

9. Giuseppe Latini, Claudio De Felice, Giuseppe Presta, et al., "In Utero Exposure to Di-(2-Ethylhexyl)phthalate and Duration of Human Pregnancy," *Environmental Health Perspectives* 111, no. 14 (December 2003): 1783–85, https://www.researchgate.net/publication/9028566_In_Utero_Exposure_to_Di-2-ethylhexylphthalate_and_Duration_of_Human_Pregnancy.

10. J. M. Braun, S. Sathyanarayana, and R. Hauser, "Phthalate Exposure and Children's Health," *Current Opinion in Pediatrics* 25, no. 2 (April 2013): 247–54.

11. F. Maranghi, R. Tassinari, V. Lagatta, et al., "Effects of the Food Contaminant Semicarbazide Following Oral Administration in Juvenile Sprague-Dawley Rats," *Food and Chemical Toxicology* 47, no. 2 (February 2009): 472–79, https://www.sciencedirect.com/science/article/pii/S0278691508006753#!.

12. Francesca Maranghi, Roberta Tassinari, Daniele Marcoccia, et al., "The Food Contaminant Semicarbazide Acts as an Endocrine Disrupter: Evidence from an Integrated *In Vivo*/*In Vitro* Approach," *Chemico-Biological Interactions* 183, no. 1 (January 2010): 40–48, https://www.sciencedirect.com/science/article/pii/S0009279709003974.

13. "EFSA Publishes Further Evaluation on Semicarbazide in Food," European Food Safety Authority, July 1, 2005, https://www.efsa.europa.eu/en/press/news/050701.

14. Mike Blake, "That Chemical Subway Ditched? McDonald's, Wendy's Use It Too," NBC News, February 7, 2014,

https://www.nbcnews.com/business/consumer/chemical-subway-ditched-mcdonalds-wendys-use-it-too-n25051.

15. Cheol-Woo Kim, Jung Hyuck Cho, Jong-Han Leem, et al., "Occupational Asthma Due to Azodicarbonamide," *Yonsei Medical Journal* 45, no. 2 (May 2004): 325–39.

16. Richard Cary, Stuart Dobson, and E. Ball, *Azodicarbonamide*, Concise International Chemical Assessment Document 16 (Geneva: World Health Organization, 1999), http://apps.who.int/iris/bitstream/handle/10665/42200/9241530162.pdf?sequence=1&isAllowed=y.

17. Joël Tassignon, Michel Vandevelde, and Michel Goldman, "Azodicarbonamide as a New T Cell Immunosuppressant: Synergy with Cyclosporin A," *Clinical Immunology* 100, no. 1 (July 2001): 24–30, https://www.sciencedirect.com/science/article/pii/S1521661601950417.

18. Kathryn J. Reid, Giovanni Santostasi, Kelly G. Baron, et al., "Timing and Intensity of Light Correlate with Body Weight in Adults," *PLOS One* (April 2, 2014): https://journals.plos.org/plosone/article?id=10.1371/journal.pone.0092251.

19. Natasha Lee, "Microorganisms Found on the Skin," DermNet NZ, August 2014, https://www.dermnetnz.org/topics/microorganisms-found-on-the-skin/.

20. Jef Akst, "Microbes of the Skin," *The Scientist*, June 13, 2014, https://mobile.the-scientist.com/article/40228/microbes-of-the-skin.

21. M. J. Blaser, M. G. Dominguez-Bello, M. Contreras,

"Distinct Cutaneous Bacterial Assemblages in a Sampling of South American Amerindians and US Residents," *Multidisciplinary Journal of Microbial Ecology* 7, no. 1 (January 2013), https://www.ncbi.nlm.nih.gov/pubmed/22895161/.

22. University of California—San Diego, "Beneficial Skin Bacteria Protect Against Skin Cancer," ScienceDaily, March 1, 2018, https://www.sciencedaily.com/releases/2018/03/180301103701.htm.

23. NIH/National Institute of Allergy and Infectious Diseases, "Bacteria Therapy for Eczema Shows Promise," Science Daily, May 3, 2018, https://www.sciencedaily.com/releases/2018/05/180503101703.htm.

24. Claudine Manach, Augustin Scalbert, and Christine Morand, "Polyphenols: Food Sources and Bioavailability," *American Journal of Clinical Nutrition* 79, no. 5 (May 2004): 727–47, https://academic.oup.com/ajcn/article/79/5/727/4690182.

25. J. Bensalem, S. Dudonné, N. Etchamendy, et al., "Polyphenols from Grape and Blueberry Improve Episodic Memory in Healthy Elderly with Lower Level of Memory Performance: A Bicentric Double-Blind, Randomized, Placebo-Controlled Clinical Study," *Journals of Gerontology* (July 2018), https://www.ncbi.nlm.nih.gov/m/pubmed/30032176/.

26. F. Afaq and S. K. Katiyar, "Polyphenols: Skin Photoprotection and Inhibition of Photocarcinogenesis,"

Mini-Reviews in Medicinal Chemistry 11, no. 14 (December 2011): 1200–1215, https://www.ncbi.nlm.nih.gov/pubmed/22070679.

27. F. Afaq, M. A. Zaid, N. Khan, et al., "Protective Effect of Pomegranate-Derived Products on UVB-Mediated Damage in Human Reconstituted Skin," *Experimental Dermatology* 18, no. 6 (June 2009): 553–61, https://www.ncbi.nlm.nih.gov/pubmed/19320737.

28. K. Kasai, M. Yoshimura, T. Koga, et al., "Effects of Oral Administration of Ellagic Acid-Rich Pomegranate Extract on Ultraviolet-Induced Pigmentation in the Human Skin," *Journal of Nutritional Science and Vitaminology (Tokyo)* 52, no. 5 (October 2006): 383–88, https://www.ncbi.nlm.nih.gov/pubmed/17190110.

29. Tomoko Tanigawa, Shigeyuki Kanazawa, Ryoko Ichibori, et al., "(+)-Catechin Protects Dermal Fibroblasts Against Oxidative Stress-Induced Apoptosis," *BMC Complementary and Alternative Medicine* 14, no. 1 (April 2014): 133, https://www.researchgate.net/publication/261514683_-Catechin_protects_dermal_fibroblasts_against_oxidative_stress-induced_apoptosis.

30. Anne-Katrin Greul, Jens-Uwe Grundmann, Felix Heinrich, et al., "Photoprotection of UV-Irradiated Human Skin: An Antioxidative Combination of Vitamins E and C, Carotenoids, Selenium and Proanthocyanidins," *Skin Pharmacology and Physiology* 15, no. 5 (September 2002): 307–15, https://www.researchgate.net/publication/11152221

_Photoprotection_of_UV-Irradiated_Human_Skin_An_Antioxidative_Combination_of_Vitamins_E_and_C_Carotenoids_Selenium_and_Proanthocyanidins.

31. Kristen Gescher, Joachim M. Kühn, Andreas Hensel, et al., "Proanthocyanidin-Enriched Extract from *Myrothamnus flabellifolia* Welw. Exerts Antiviral Activity Against Herpes Simplex Virus Type 1 by Inhibition of Viral Adsorption and Penetration," *Journal of Ethnopharmacology* 134, no. 2 (March 24, 2011): 468–74, https://www.ncbi.nlm.nih.gov/pubmed/21211557.

32. Chwan-Fwu Lin, Yann-Lii Leu, Saleh A. Al-Suwayeh, et al., "Anti-Inflammatory Activity and Percutaneous Absorption of Quercetin and Its Polymethoxylated Compound and Glycosides: The Relationships to Chemical Structures," *European Journal of Pharmaceutical Sciences* 47, no. 5 (May 2012): 857–64, https://www.sciencedirect.com/science/article/pii/S0928098712002096.

33. Jheng-Hua Huang, Chieh-Chen Huang, Jia-You Fang, et al., "Protective Effects of Myricetin Against Ultraviolet-B-Induced Damage in Human Keratinocytes," *Toxicology in Vitro* 24, no. 1 (February 2010): 21–28, https://www.sciencedirect.com/science/article/pii/S0887233309002823.

34. Shi-Hui Dong, Geping Cai, José G. Napolitano, et al., "Lipidated Steroid Saponins from *Dioscorea villosa* (Wild Yam)," *Fitoterapia* 91 (December 2013): 113–24, https://www.sciencedirect.com/science/article/pii/S0367326X13002013.

35. Tessa Moses, Kalliope K. Papadopoulou, and Anne Osbourn, "Metabolic and Functional Diversity of Saponins, Biosynthetic Intermediates and Semi-Synthetic Derivatives," *Critical Reviews in Biochemistry and Molecular Biology* 49, no. 6 (October 6, 2014): 439–62, https://www.tandfonline.com/doi/full/10.3109/10409238.2014953628.

36. Vamshi K. Manda, Bharathi Avula, Zulfiqar Ali, et al., "Characterization of *In Vitro* ADME Properties of Diosgenin and Dioscin from *Dioscorea villosa*," *Planta Medica* 79, no. 15 (October 2013): 1421–28, https://www.ncbi.nlm.nih.gov/pmc/articles/PMC5592720/.

37. Yayoi Tada, Naoko Kanda, Akinori Haratake, et al., "Novel Effects of Diosgenin on Skin Aging," *Steroids* 74, no. 6 (June 2009): 504–11, https://www.sciencedirect.com/science/article/pii/S0039128X09000233.

38. Jongsung Lee, Kwangseon Jung, Yeong Shik Kim, et al., "Diosgenin Inhibits Melanogenesis Through the Activation of Phosphatidylinositol-3-Kinase Pathway (PI3K) Signaling," *Life Sciences* 81, no. 3 (June 27, 2007): 249–54, https://www.sciencedirect.com/science/article/pii/S002432050700375X.

39. Ivana Binic, Viktor Lazarevic, Milanka Ljubenovic, et al., "Skin Ageing: Natural Weapons and Strategies," *Evidence-Based Complementary and Alternative Medicine* 2013 (2013): 827248, https://www.researchgate.net/publication/235691895_Skin_Ageing_Natural_Weapons_and_Strategies.

40. Ibid.
41. Manda et al., "Characterization of *In Vitro* ADME Properties of Diosgenin and Dioscin from *Dioscorea villosa*."
42. Stacie E. Geller and Laura Studee, "Botanical and Dietary Supplements for Menopausal Symptoms: What Works, What Does Not," *Journal of Women's Health* 14, no. 7 (October 2005): 634–49, https://www.researchgate.net/publication/7583156_Botanical_and_Dietary_Supplements_for_Menopausal_Symptoms_What_Works_What_Does_Not.
43. Manda et al., "Characterization of *In Vitro* ADME Properties of Diosgenin and Dioscin from *Dioscorea villosa*."
44. Ibid.

Chapter 8: The Longevity Paradox Foods

1. Embriette Hyde, "What Does a Three-Day Dietary Cleanse Do to Your Gut Microbiome?," American Gut, February 29, 2016, http://americangut.org/what-does-a-three-day-dietary-cleanse-do-to-your-gut-microbiome/.
2. Delfin Rodriguez-Leyva, Chantal M. C. Bassett, Richelle McCullough, et al., "The Cardiovascular Effects of Flaxseed and Its Omega-3 Fatty Acid, Alpha-Linolenic Acid," *Canadian Journal of Cardiology* 26, no. 9 (November 2010): 489–96, https://www.ncbi.nlm.nih.gov/pmc/articles/PMC2989356/.

3. Penn State, "Like It or Not: Broccoli May Be Good for the Gut," ScienceDaily, October 12, 2017, https://www.sciencedaily.com/releases/2017/10/171012151754.htm.
4. J. Mercola, "Why Aged Cheese and Mushrooms Are So Good for Your Heart (and Make You Live Longer Too)," The Science of Eating, 2016, http://thescienceofeating.com/2017/08/17/aged-cheese-mushrooms-good-heart-make-live-longer/.
5. Tobias Eisenberg, Mahmoud Abdellatif, Frank Madeo, et al., "Cardioprotection and Lifespan Extension by the Natural Polyamine Spermidine," *Nature Medicine* 22 (November 2016): 1428–38, https://www.nature.com/articles/nm.4222.
6. Charles N. Serhan, Song Hong, Karsten Gronert, et al., "Resolvins: A Family of Bioactive Products of Omega-3 Fatty Acid Transformation Circuits Initiated by Aspirin Treatment That Counter Proinflammation Signals," *Journal of Experimental Medicine* 196, no. 8 (November 2002): 1025–37, https://www.researchgate.net/publication/11071436_Resolvins_A_Family_of_Bioactive_Products_of_Omega-3_Fatty_Acid_Transformation_Circuits_Initiated_by_Aspirin_Treatment_that_Counter_Proinflammation_Signals.
7. Alexander Obrosov, Lawrence J. Coppey, Hanna Shevalye, et al., "Effect of Fish Oil vs. Resolvin D1, E1, Methyl Esters of Resolvins D1 or D2 on Diabetic Peripheral Neu-

ropathy," *Journal of Neurology and Neurophysiology* 8, no. 6 (2017): 453, https://www.ncbi.nlm.nih.gov/pmc/articles/PMC5800519/.

8. Ze-Jian Wang, Cui-Ling Liang, Guang-Mei Li, et al., "Neuroprotective Effects of Arachidonic Acid Against Oxidative Stress on Rat Hippocampal Slices," *Chemico-Biological Interactions* 163, no. 3 (November 7, 2006): 207–17, https://www.sciencedirect.com/science/article/pii/S0009279706002122.

9. Jyrki K. Virtanen, Jason H. Y. Wu, Sari Voutilainen, et al., "Serum n-6 Polyunsaturated Fatty Acids and Risk of Death: The Kuopio Ischaemic Heart Disease Risk Factor Study," *American Journal of Clinical Nutrition* 107, no. 3 (March 2018): 427–35, https://www.researchgate.net/publication/323967337_Serum_n-6_polyunsaturated_fatty_acids_and_risk_of_death_The_Kuopio_Ischaemic_Heart_Disease_Risk_Factor_Study.

10. Michael D. Roberts, Mike Iosia, Chad M. Kerksick, et, al., "Effects of Arachidonic Acid Supplementation on Training Adaptations in Resistance-Trained Males," *Journal of the International Society of Sports Nutrition* 4, no. 21 (November 2007): https://jissn.biomedcentral.com/articles/10.1186/1550-2783-4-21.

11. James V. Pottala, Kristine Yaffe, Jennifer G. Robinson, et al., "Higher RBC EPA + DHA Corresponds with Larger Total Brain and Hippocampal Volumes: WHIMS-MRI Study," *Neurology* 82, no. 5 (January 2014): 435–42, https://www.researchgate.net/publication/259877890_Higher_RBC

_EPA_DHA_corresponds_with_larger_total_brain_and _hippocampal_volumes_WHIMS-MRI_study.

12. Alexandra J. Richardson, Jennifer R. Burton, Richard P. Sewell, et al., "Docosahexaenoic Acid for Reading, Cognition and Behavior in Children Aged 7–9 Years: A Randomized, Controlled Trial (The DOLAB Study)," *PLOS One*, September 6, 2012, https://journals.plos.org/plosone /article?id=10.1371/journal.pone.0043909.
13. University of Massachusetts Lowell, "Omega-3s Help Keep Kids Out of Trouble: How Diet, Biology Can Prevent Bad, Even Criminal Behavior," ScienceDaily, July 24, 2018, https://www.sciencedaily.com/releases /2018/07/180724174322.htm.
14. Stephanie Liou, "Brain-Deprived Neurotrophic Factor (BDNF)," Huntington's Outreach Project for Education at Stanford, June 26, 2010, http://web.stanford.edu/group /hopes/cgi-bin/hopes_test/brain-derived-neurotrophic -factor-bdnf/#how-does-bdnf-work.
15. Astrid Nehlig, "The Neuroprotective Effects of Cocoa Flavanol and Its Influence on Cognitive Performance," *British Journal of Clinical Pharmacology* 75, no. 2 (March 2013): 716–27, https://bpspubs.onlinelibrary.wiley.com/doi/abs /10.1111/j.1365-2125.2012.04378.x.
16. Dayong Wu, Junpeng Wang, Munkyong Pae, et al., "Green Tea EGCG, T Cells, and T Cell–Mediated Autoimmune Diseases," *Molecular Aspects of Medicine* 33, no. 1 (February 2012): 107–18, https://www.sciencedirect.com /science/article/pii/S0098299711000458.

17. Yan Hou, Wanfang Shao, Rong Xiao, et al., "Pu-erh Tea Aqueous Extracts Lower Atherosclerotic Risk Factors in a Rat Hyperlipidemia Model," *Expermental Gerontology* 44, nos. 6–7 (June–July 2009): 434–39, https://www.sciencedirect.com/science/article/pii/S053155650900062X.

18. Ida M. Klang, Birgit Schilling, Dylan J. Sorensen, et al., "Iron Promotes Protein Insolubility and Aging in *C. elegans*," *Aging (Albany NY)* 6, no. 11 (November 2014): 975–88, http://europepmc.org/articles/PMC4276790/.

19. Michael Day, "Give Blood, Live Longer," *New Scientist*, October 17, 1998, https://www.newscientist.com/article/mg16021562-500-give-blood-live-longer/.

20. M. B. Abou-Donia, E. M. El-Masry, A. A. Abdel-Rahman, et al., "Splenda Alters Gut Microflora and Increases Intestinal P-Glycoprotein and Cytochrome P-450 in Male Rats," *Journal of Toxicology and Environmental Health* 7, no. 21 (2008): https://www.ncbi.nlm.nih.gov/m/pubmed/18800291/.

21. M. Y. Pepino, C. D. Tiemann, B. W. Patterson, et al., "Sucralose Affects Glycemic and Hormonal Responses to an Oral Glucose Load," *Diabetes Care* 36, no. 9 (September 2013): 2530–35, https://www.ncbi.nlm.nih.gov/m/pubmed/23633524/.

22. Susan S. Schiffman and Kristina I. Rother, "Sucralose, a Synthetic Organochlorine Sweetener: Overview of Biological Issues," *Journal of Toxicology and Environmental Health* 16, no. 7 (2013): 399–451, https://www.tandfonline.com/doi/pdf/10.1080/10937404.2013.842523.

23. Qiang Wang, Lu-Gang Yu, Barry J. Campbell, et al., "Identification of Intact Peanut Lectin in Peripheral Venous Blood," *Lancet* 352, no. 9143 (December 5, 1998): 1831–32, https://www.sciencedirect.com/science/article/pii/S0140673605798949.

Chapter 9: The Longevity Paradox Meal Plan

1. American Psychological Association, "You're Only as Old as You Think and Do: Increased Control, Physical Activity Lower Subjective Age in Older Adults, Research Says," ScienceDaily, August 9, 2018, https://www.sciencedaily.com/releases/2018/08/180809141122.htm.
2. In Young Choi, Laura Piccio, Patra Childress, et al., "A Diet Mimicking Fasting Promotes Regeneration and Reduces Autoimmunity and Multiple Sclerosis Symptoms," *Cell Reports* 15, no. 10 (June 2016), 2136–46, http://www.cell.com/cell-reports/fulltext/S2211-1247(16)30576-9.
3. Embriette Hyde, "What Does a Three-Day Dietary Cleanse Do to Your Gut Microbiome?," American Gut, February 29, 2016, http://americangut.org/what-does-a-three-day-dietary-cleanse-do-to-your-gut-microbiome/.
4. Ernst J. Drenick, Lia C. Alvarez, Gabor C. Tamasi, et al., "Resistance to Symptomatic Insulin Reactions after Fasting," *Journal of Clinical Investigation* 51, no. 10 (October 1972): 2757–62, https://www.researchgate.net/publication/18111482_Resistance_to_Symptomatic_Insulin_Reactions_after_Fasting.

5. NIH/National Institute on Aging, "Longer Daily Fasting Times Improve Health and Longevity in Mice: Benefits Seen Regardless of Calorie Intake, Diet Composition in New Study," ScienceDaily, September 6, 2018, https://www.sciencedaily.com/releases/2018/09/180906123305.htm.
6. Dagmar E. Ehrnhoefer, Dale D. O. Martin, Mandi E. Schmidt, et al., "Preventing Mutant Huntingtin Proteolysis and Intermittent Fasting Promote Autophagy in Models of Huntington Disease," *Acta Neuropathologica Communications* 6, no. 16 (March 2018): https://actaneurocomms.biomedcentral.com/articles/10.1186/s40478-018-0518-0.
7. Prashant K. Nighot, Chien-An Andy Hu, and Thomas Y. Ma, "Autophagy Enhances Intestinal Epithelial Tight Junction Barrier Function by Targeting Claudin-2 Protein Degradation," *Journal of Biological Chemistry* 290, no. 11 (March 2015): 77234–46, http://m.jbc.org/content/290/11/7234.full.
8. Elitsa Ananieva, "Targeting Amino Acid Metabolism in Cancer Growth and Anti-Tumor Immune Response," *World Journal of Biological Chemistry* 6, no. 4 (November 2015): 281–89, http://www.ncbi.nlm.nih.gov/pmc/articles/PMC4657121/.

Chapter 10: The Longevity Paradox Lifestyle Plan

1. Y. Koh and J. Park, "Cell Adhesion Molecules and Exercise," *Journal of Inflammation Research* 11 (July 2018): 297–306, https://www.dovepress.com/articles.php?article_id=39476.
2. Changin Oh, Kunkyu Lee, Yeotaek Cheong, et al., "Comparison of the Oral Microbiomes of Canines and Their Owners Using Next-Generation Sequencing," *PLOS One* 10, no. 7 (July 2015): http://journals.plos.org/plosone/article?id=10.1371/journal.pone.0131468.
3. University of British Columbia, Okanagan Campus, "Short Bursts of Intense Exercise Are a HIIT, Even with Less Active People: Participants Find High-Intensity Interval Exercise as Enjoyable as Traditional Exercise," ScienceDaily, May 24, 2018, https://www.sciencedaily.com/releases/2018/05/180524141625.htm.
4. Elsevier, "Frequent Sauna Bathing Has Many Health Benefits," ScienceDaily, August 1, 2018, https://www.sciencedaily.com/releases/2018/08/180801131602.htm.
5. Bi Zhang, Jianke Gong, Wenyuan Zhang, et al., "Brain-Gut Communications via Distinct Neuroendocrine Signals Bidirectionally Regulate Longevity in *C. elegans*," *Genes and Development* 32, nos. 3–4 (January 2018): http://m.genesdev.cshlp.org/content/32/3-4/258.
6. NIH/National Institute on Alcohol Abuse and Alcoholism, "Lack of Sleep May Be Linked to Risk Factor for

Alzheimer's Disease: Preliminary Study Shows Increased Levels of Beta-Amyloid," ScienceDaily, April 13, 2018, https://www.sciencedaily.com/releases/2018/04/180413155301.htm.

7. K. Spiegel, E. Tasali, P. Penev, et al., "Brief Communication: Sleep Curtailment in Healthy Young Men Is Associated with Decreased Leptin Levels, Elevated Ghrelin Levels, and Increased Hunger and Appetite," *Annals of Internal Medicine* 141, no. 11 (December 2004): 846–50, https://www.ncbi.nlm.nih.gov/pubmed/15583226.
8. Technical University of Munich (TUM), "Effect of Genetic Factors on Nutrition: The Genes Are Not to Blame," ScienceDaily, July 20, 2018, https://www.sciencedaily.com/releases/2018/07/180720092515.htm.
9. Joe Alcock, Carlo C. Maley, and C. Athena Aktipis, "Is Eating Behavior Manipulated by the Gastrointestinal Microbiota? Evolutionary Pressures and Potential Mechanisms," *Bioessays* 36, no. 10 (October 2014): 940–49, https://www.ncbi.nlm.nih.gov/pmc/articles/PMC4270213/.
10. N. A. Christakis and J. H. Fowler, "The Spread of Obesity in a Large Social Network over 32 Years," *New England Journal of Medicine* 357, no.4 (July 2007): 370–79, https://www.ncbi.nlm.nih.gov/pubmed/17652652/.
11. Remco Kort, Martien Caspers, Astrid van de Graaf, et al., "Shaping the Oral Microbiota Through Intimate Kissing," *Microbiome* 2, no. 41 (November 2014): https://microbiomejournal.biomedcentral.com/articles/10.1186/2049-2618-2-41.

Chapter 11: Longevity Paradox Supplement Recommendations

1. American Heart Association, "Diet, 'Anti-Aging' Supplements May Help Reverse Blood Vessel Abnormality," ScienceDaily, May 1, 2013, https://www.sciencedaily.com/releases/2013/05/130501193127.htm.
2. E. L. Stenblom, E. Egecioglu, M. Landin-Olsson, et al., "Consumption of Thylakoid-Rich Spinach Extract Reduces Hunger, Increases Satiety and Reduces Cravings for Palatable Food in Overweight Women," *Appetite* 91 (2015): 209–19, http://www.ncbi.nlm.nih.gov/m/pubmed/25895695/.
3. D. Thomas, "The Mineral Depletion of Foods Available to Us as a Nation (1940–2002)—A Review of the 6th Edition of McCance and Widdowson," *Nutrition and Health* 19, nos. 1–2 (2007): 21–55, http://www.ncbi.nlm.nih.gov/m/pubmed/18309763/.

About the Author

Steven R. Gundry, MD, is a cum laude graduate of Yale University, with special honors in human biological and social evolution. After graduating Alpha Omega Alpha from the Medical College of Georgia, Dr. Gundry completed residencies in general surgery and cardiothoracic surgery at the University of Michigan and served as a clinical associate at the National Institutes of Health. He invented devices that reverse the cell death seen in heart attacks; variations of these devices became the Medtronic Gundry Retrograde Cardioplegia Cannula, the most widely used device of its kind worldwide to protect the heart during open-heart surgery. After completing a fellowship in congenital heart surgery at the Hospital for Sick Children, Great Ormond Street, in London, and spending two years as a professor at the

University of Maryland School of Medicine, Dr. Gundry was recruited as professor and chairman of cardiothoracic surgery at Loma Linda University School of Medicine.

During his tenure at Loma Linda, Dr. Gundry pioneered the field of xenotransplantation, the study of how the immune system and blood vessel proteins of one species react to the transplanted heart of a foreign species. He was one of the original twenty investigators of the first FDA-approved implantable left ventricular assist device. Dr. Gundry is the inventor of the Gundry Ministernotomy, the most widely used minimally invasive surgical technique to operate on the aortic valve; the Gundry Lateral Tunnel, a living tissue that can rebuild parts of the heart in children with severe congenital heart malformations; and the Skoosh Venous Cannula, the most widely used cannula in minimally invasive heart operations.

As a consultant to Computer Motion (now Intuitive Surgical), Dr. Gundry was one of the fathers of robotic heart surgery. He received early FDA approval for robotic-assisted minimally invasive surgery for coronary artery bypass and mitral valve operations. He holds patents on connecting blood vessels and coronary artery bypasses without the need for sutures, as well as on repairing the mitral valve without the need for sutures and the heart-lung machine.

Dr. Gundry has served on the Board of Directors of the American Society of Artificial Internal Organs and was a founding board member and treasurer of the International Society of Minimally Invasive Cardiothoracic Surgery. He also served two successive terms as president of the Board of Directors of the American Heart Association, Desert Division. Dr. Gundry has been elected a fellow of the American College of Surgeons, the American College of Cardiology, the American Surgical Association, the American Academy of Pediatrics, and the College of Chest Physicians. He has served numerous times as an abstract reviewer for the American Heart Association annual meetings. The author of more than three hundred articles, chapters, and abstracts in peer-reviewed journals on surgical, immunologic, genetic, nutritional, and lipid investigations, he has also operated in more than thirty countries, including on multiple charitable missions.

In 2000, inspired by the stunning reversal of coronary artery disease in an "inoperable" patient by using a combination of dietary changes and nutriceutical supplements, Dr. Gundry changed the arc of his career. An obese chronic diet failure himself, he adapted his Yale University thesis to design a diet based on evolutionary coding and the interaction of our ancestral microbiome, genes, and environment. Following this program enabled him to reverse his own numerous medical problems. In the process, he effortlessly lost seventy pounds and has kept them off for seventeen years. These discoveries led him to establish the International Heart and Lung Institute—and, as part of it, the Center for Restorative Medicine—in Palm Springs and Santa Barbara, California. There he has devoted his research and clinical practice to the dietary and nutriceutical reversal of most diseases, including heart disease, diabetes, autoimmune disease, cancer, arthritis, kidney failure, and neurological conditions such as dementia and Alzheimer's disease, using sophisticated blood tests and blood flow measurements to maximize his patients' health span and longevity.

This research resulted in the publication of his bestselling first book, *Dr. Gundry's Diet Evolution: Turn Off the Genes That Are Killing You and Your Waistline*, in 2008. Following up on the success of that book, he has become one of the world's authorities on the human microbiome and the interaction between the gut, the foods we ingest, the products we use, and our physical and mental health and well-being. In recent years, more than 50 percent of his practice has been devoted to the reversal of challenging autoimmune conditions in patients referred to him by health professionals around the world. These findings resulted in the publication of the *New York Times* bestseller *The Plant Paradox* and *The Plant Paradox Cookbook*, as well as *The Plant Paradox: Quick and Easy*, recently released. *The Plant Paradox* has been translated into over twenty-five languages and has prompted worldwide interest in a lectin-free diet.

Dr. Gundry has been named to America's Top Doctors for twenty-one years in a row by Castle Connolly, the independent physician rating company; to *Palm Springs Life* Top Doctors for fifteen years in a row; and to *Los Angeles Magazine*'s Top Doctors for the last six years.

Dr. Gundry is the creator of the nutritional guidelines for the Six Senses Resorts and Spas worldwide and a senior scientific advisor to Pegasus Capital Advisors. He has been invited to lecture at both the Stanford and MIT Brain Summit meetings on the impact of the gut on brain health and its deterioration. In 2016, he founded GundryMD, his own line of nutriceutical and skin-care supplements.

Dr. Gundry's wife, Penny, and their dogs, Pearl, Minnie, and George, live in Palm Springs and Montecito California. His grown daughters, Elizabeth and Melissa, their husbands, Tim and Ray, and their children, Sophie and Oliver, live nearby.